WILLIAM F. MAAG LIBRARY
YOUNGSTOWN STATE UNIVERSITY

Advances in
INORGANIC CHEMISTRY

Volume 31

Advances in
INORGANIC CHEMISTRY

EDITORS

H. J. EMELÉUS

A. G. SHARPE

*University Chemical Laboratory
Cambridge, England*

VOLUME 31

1987

ACADEMIC PRESS, INC.
Harcourt Brace Jovanovich, Publishers
Orlando San Diego New York Austin
Boston London Sydney Tokyo Toronto

COPYRIGHT © 1987 BY ACADEMIC PRESS, INC.
ALL RIGHTS RESERVED.
NO PART OF THIS PUBLICATION MAY BE REPRODUCED OR
TRANSMITTED IN ANY FORM OR BY ANY MEANS, ELECTRONIC
OR MECHANICAL, INCLUDING PHOTOCOPY, RECORDING, OR
ANY INFORMATION STORAGE AND RETRIEVAL SYSTEM, WITHOUT
PERMISSION IN WRITING FROM THE PUBLISHER.

ACADEMIC PRESS, INC.
Orlando, Florida 32887

United Kingdom Edition published by
ACADEMIC PRESS INC. (LONDON) LTD.
24–28 Oval Road, London NW1 7DX

LIBRARY OF CONGRESS CATALOG CARD NUMBER: 59-7692

ISBN 0–12–023631-1 (alk. paper)

PRINTED IN THE UNITED STATES OF AMERICA

87 88 89 90   9 8 7 6 5 4 3 2 1

CONTENTS

Preparation and Purification of Actinide Metals

J. C. Spirlet, J. R. Peterson, and L. B. Asprey

I.	Introduction	1
II.	Preparation of Actinide Metals	4
III.	Purification (Refining) of Actinide Metals	11
IV.	Growth of Single Crystals of Actinide Metals	14
V.	Specifics of Actinide Metal Preparation	15
VI.	Some Physical Properties of Actinide Metals	36
	References	37

Astatine: Its Organonuclear Chemistry and Biomedical Applications

I. Brown

I.	Introduction	43
II.	Isotopes: Production, Extraction, and Identification	44
III.	Synthetic Organic Radiochemistry	49
IV.	Biological Behavior	77
V.	Biomedical Applications—Cancer Therapy	79
	References	83

Polysulfide Complexes of Metals

A. Müller and E. Diemann

I.	Introduction	89
II.	Polysulfide Ions and Solutions	90
III.	Survey of Compound Types	91
IV.	Syntheses	103
V.	Reactions of Coordinated Ligands	106

VI.	Spectroscopic Properties	109
VII.	Some Structural Features and Chemical Bonding	111
	References	116

IMINOBORANES

Peter Paetzold

I.	Introduction	123
II.	Formation of Iminoboranes	124
III.	Structure of Iminoboranes	133
IV.	Oligomerization of Iminoboranes	141
V.	Polar Additions to Iminoboranes	150
VI.	Iminoboranes as Components in Cycloaddition Reactions	159
VII.	Iminoboranes in the Coordination Sphere of Transition Metals	165
	References	168

Synthesis and Reactions of Phosphorus-Rich Silylphosphanes

G. Fritz

I.	Introduction	171
II.	Formation of $P_7(SiMe_3)_3$	173
III.	Formation of Cyclic Silylphosphanes	175
IV.	Synthesis and Reactions of Silylated Triphosphanes and Triphosphides	188
V.	Synthesis of Silylated Tri- and Tetraphosphanes via Lithiated Diphosphanes, and Their Reactions	194
VI.	Silylated Cyclotetraphosphanes	198
VII.	Reactions of Silylated Triphosphanes and Cyclotetraphosphanes with Lithium Alkyls	199
	References	212

Index 215

PREPARATION AND PURIFICATION OF ACTINIDE METALS

J. C. SPIRLET,* J. R. PETERSON,** and L. B. ASPREY[†]

*Commission of the European Communities,
Joint Research Centre, Karlsruhe Establishment,
European Institute for Transuranium Elements,
D-7500 Karlsruhe 1, Federal Republic of Germany
**Department of Chemistry, University of Tennessee,
Knoxville, Tennessee 37996-1600 and Transuranium Research Laboratory,
Chemistry Division, Oak Ridge National Laboratory,
Oak Ridge, Tennessee 37831
[†]Isotope and Nuclear Chemistry Division,
Los Alamos National Laboratory,
University of California,
Los Alamos, New Mexico 87545

I. Introduction

The first actinide metals to be prepared were those of the three members of the actinide series present in nature in macro amounts, namely, thorium (Th), protactinium (Pa), and uranium (U). Until the discovery of neptunium (Np) and plutonium (Pu) and the subsequent manufacture of milligram amounts of these metals during the hectic World War II years (i.e., the early 1940s), no other actinide element was known. The demand for Pu metal for military purposes resulted in rapid development of preparative methods and considerable study of the chemical and physical properties of the other actinide metals in order to obtain basic knowledge of these unusual metallic elements.

These early studies were carried out on metals of typically 90–99% purity, which sufficed to determine at least their gross properties. During the 1960s, interest diminished somewhat in actinide metallurgy due in part to the increasing use of ceramic rather than metallic fuel elements in nuclear reactors. The bulk of actinide metal research was for secret military purposes and only a fraction of the fundamental research was published.

Interest was rekindled during the 1970s due to advances in solid-state physics and a growing realization of the unusual properties of the lighter actinide metals due to the behavior of their $5f$ electrons.

Demand for high-purity metals, on the order of 99.9 to 99.999 atomic percent (at %) purity, resulted in renewed study of actinide metallurgy with emphasis on improved methods for preparation, refining, and growth of single crystals. Increased interest in the properties of highly symmetric binary compounds of actinides such as chalcogenides, pnictides, carbides, etc., which are made by direct combination of the elements, demanded yet further development of new and improved synthesis and crystal-growing techniques.

The actual situation with regard to the purity of most of the actinide metals is far from ideal. Only thorium (99), uranium (11, 17), neptunium (20), and plutonium (60) have been produced at a purity ≥ 99.9 at %. Due to the many grams required for preparation and for accurate analysis, it is probable that these abundant and relatively inexpensive elements (Table I) are the only ones whose metals can be prepared and refined to give such high purities, and whose purity can be verified by accurate analysis. The purity levels achieved for some of the actinide metals are listed in Table II. For actinium (Ac), berkelium (Bk), californium (Cf),

TABLE I

AVAILABILITY AND PRICE OF SELECTED ACTINIDE ISOTOPES

Element	Isotope	Available quantity	Price ($/g)	Limited handling quantity[a]	Hazardous radiation
Ac	227	mg	—	<mg	α (21.8 years)
					γ (daughters)
Th	232	kg	—	—	—
Pa	231	g	50,000	g	γ (daughters)
U	Natural	kg	20[b]	kg	—
	235	g	150	kg	n
	238	kg	100	—	—
Np	237	kg	500[b]	kg	γ (daughters)
Pu	238	g	5,000	g	α (87.7 years)
	239	kg	500[b]	250 g	γ (daughters), n
	242	g	15,000	10 g	n
Am	241	kg	2,000	mg	γ
	243	g	100,000	mg	γ (daughters)
Cm	244	g	100,000	mg	n,γ
	248	mg	[c]	mg	n
Bk	249	mg	[c]	mg	n,γ (daughters)
Cf	249	mg	[c]	mg	α (351 years); n,γ

[a] Maximum quantity of the isotope which can be handled in a standard gloved box without special shielding.
[b] Electrorefined metal.
[c] Not available commercially.

TABLE II

PURITY OF SOME ACTINIDE METALS

Element	Purity (atom %)
Thorium	99.99
Protactinium	99.7
Uranium	99.9
Neptunium	99.9
Plutonium	99.9
Americium	99.5
Curium	99.5

and einsteinium (Es) metals, only limited data on metallic impurities exist, so no values are given in Table II. The values for thorium, uranium, neptunium, and plutonium are very reliable. Complete analysis of the impurities in these metals requires at least 5 g of metal and is very expensive. Metals of ultrahigh purity are analyzed indirectly by determining the ratio of their electrical resistivity at 300 K to that at 4.2 K, a measurement which relates only to the total quantity of impurities (see footnote 1, Section III,D).

Improvements in preparative techniques are expected to increase the quality of the rarer metals, but complete analysis to establish the purity will probably remain unavailable and such work will be very time consuming and costly.

The preparation, refining, and growth of single crystals of actinide metals is a very difficult task for a number of reasons. One is that the physical properties of the metals vary enormously. They range from those typical of the transition elements for the lighter actinides to those typical of rare earth metals for Am, Cm, and beyond. Their vapor pressures vary over 10 orders of magnitude, their melting points cover the range from 913 to 2028 K, and they exhibit from none to five phase transitions (some with large density differences) with increasing temperature or pressure.

Another difficulty arises from the chemical properties of the actinide metals. They are chemically reactive, rapidly corroded by moist air, pyrophoric, and, when in the molten state, dissolve common crucible materials. The radioactivity of short-lived isotopes of Am and Cm makes their long-term storage difficult; small amounts can be stored successfully under ultrahigh vacuum. Large amounts of isotopes such as ^{238}Pu with a $\tau_{1/2}$ of only 87.7 years are best stored under a pure inert gas such as argon in order to remove the heat generated by the radioactive decay. To further complicate matters, the radioactivity and toxicity of these metals make it mandatory to carry out all procedures in gloved boxes or hot cells, whether the preparation is on a microgram or multigram scale, making contamination of the product metal more difficult to avoid. Thus, sophisticated techniques are necessary to prepare, refine, and grow single crystals of actinide metals. These techniques must be compatible with highly radioactive and chemically reactive materials. Due to the varying availability and cost of the actinide isotopes (Table I), these techniques must be applicable to submilligram up to multigram quantities of actinides. By far the biggest difficulty associated with the preparation of high-purity metals is determining the purity of the product metal. Adequate analytical methods are usually not available and must be developed.

This article presents a general discussion of actinide metallurgy, including advanced methods such as levitation melting and chemical vapor-phase reactions. A section on purification of actinide metals by a variety of techniques is included. Finally, an element-by-element discussion is given of the most satisfactory metallurgical preparation for each individual element: actinium (included for completeness even though not an actinide element), thorium, protactinium, uranium, neptunium, plutonium, americium, curium, berkelium, californium, and einsteinium.

The transeinsteinium actinides, fermium (Fm), mendelevium (Md), nobelium (No), and lawrencium (Lr), are not available in weighable ($>\mu$g) quantities, so these elements are unknown in the condensed bulk phase and only a few studies of their physicochemical behavior have been reported. Neutral atoms of Fm have been studied by atomic beam magnetic resonance (47). Thermochromatography on titanium and molybdenum columns has been employed to characterize some metallic state properties of Fm and Md (61). This article will not deal with the preparation of these transeinsteinium metals.

II. Preparation of Actinide Metals

A. METALLOTHERMIC REDUCTION OF ACTINIDE HALIDES OR OXIDES

Metallothermic reduction of an actinide halide was the first method applied to the preparation of an actinide metal. Initially, actinide chlorides were reduced by alkali metals, but then actinide fluorides, which are much less hygroscopic than the chlorides, were more generally used. On the milligram scale, the anhydrous fluorides are reduced by alkali (usually Li) or alkaline earth (usually Ba) metal vapor in a covered tantalum or ceramic crucible. The excess reductant and the fluoride salt by-product are eliminated by vaporization through an effusion hole in the crucible cover.

The vapor pressures at 1473 K of a few of the actinide elements and other materials of interest are given in Table III. All of the actinide (An) elements through einsteinium can be obtained by this process:

$$AnF_3 + 3Li \longrightarrow 3LiF + An \quad \text{or} \quad AnF_4 + 4Li \longrightarrow 4LiF + An$$

although significant losses of Am, Cf, and Es metals result because of their high volatilities (Fig. 1). A review of this topic has been published (53).

This method has also been successfully adapted to the pilot-plant and industrial-scale production of Th, U, Np, and Pu metals. In these cases, calcium metal is preferred as the reductant, and the reaction is:

$$AnF_4 + 2Ca \longrightarrow 2CaF_2 + An$$

This Ca reduction technique is used widely to produce commercially available actinide metals. However, this method is not well suited to the preparation on the laboratory scale of pure (>99.9 at %) actinide

TABLE III

APPROXIMATE VAPOR PRESSURES OF SELECTED ELEMENTS AND COMPOUNDS

Metal or compound	Vapor pressure (Pa)[a]
Zn	1.3×10^6
Mg	$>1.3 \times 10^5$
Ca	1.2×10^4
Am	1.3×10
Pu	1.3×10^{-2}
La	1.3×10^{-4}
U	1.3×10^{-6}
Pt	1.3×10^{-8}
Th	1.3×10^{-8}
$MgCl_2$	2.7×10^4
PuF_4	1.3×10^2
$CaCl_2$	1.3×10^{-1}
CaF_2	1.3×10^{-4}

[a] At 1473 K.

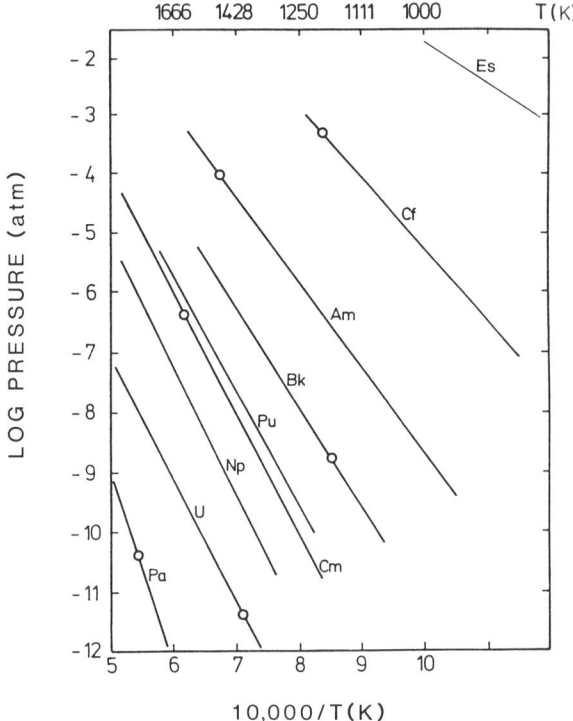

FIG. 1. Vapor pressures and some melting points of actinide metals Pa–Es. ○, melting point.

metals for solid-state physics investigations. Preparation of anhydrous actinide fluorides by treatment of the corresponding oxides with HF or F_2 gas is a source of operating difficulties. The personnel exposure to neutrons resulting from the (α,n) reaction on fluorine nuclei is high when working on the multigram scale of operation. Yield of the metallothermic reduction is somewhat irregular, and recovery of the actinide metal from the crucible and/or slag (CaF_2) can be difficult. All nonvolatile impurities present in the reductant metal as well as corrosion products from the crucible frequently end up in the product metal. If great care is taken to prepare pure anhydrous halides and to refine the reductants by successive distillation, the metal product can be >99 at % pure. Further refining steps are necessary to obtain metal of higher purity.

The fluoride reduction technique is being replaced in production plants by the metallothermic reduction of an oxide (24). Direct oxide reduction (DOR) is used for preparation of U, Np, and Pu on the kilogram or larger scale (24). The dioxide is added to molten $CaCl_2$ in the presence of excess Ca metal. The molten solvent serves to dissolve the CaO produced in the reduction reaction:

$$AnO_2 + 2Ca(xs) \longrightarrow 2CaO \text{ (soln)} + An$$

Excellent quality metal, comparable to that from the halide reduction, can be prepared by this technique. A big advantage is that no neutrons are present from (α,n) reactions on fluorine nuclei, in marked contrast to the case with actinide fluorides.

B. Metallothermic Reduction of Actinide Oxides Followed by Distillation

The metallothermic reduction of an oxide is a useful preparative method for an actinide metal when macro quantities of the actinide are available. A mixture of the actinide oxide and reductant metal is heated in vacuum at a temperature which allows rapid vaporization of the actinide metal, leaving behind an oxide of the reductant metal and the excess reductant metal, in accord with the following equations:

$$An_2O_3 + 2La \longrightarrow La_2O_3 + 2An \quad \text{and} \quad 2An_2O_3 + 3Th \longrightarrow 3ThO_2 + 4An$$

The reductant metal must have the following properties: (1) the free energy of formation of the oxide of the reductant has to be more negative than that of the actinide oxide; and (2) the vapor pressure of the reductant metal needs to be smaller by several orders of magnitude than that of the actinide metal. This difference in vapor pressure should be at least five orders of magnitude to keep the contamination level of the co-evaporated reductant metal in the product actinide metal below the 10 ppm level.

Americium, californium, and einsteinium oxides have been reduced by lanthanum metal, whereas thorium has been used as the reductant metal to prepare actinium, plutonium, and curium metals from their respective oxides. Berkelium metal could also be prepared by Th reduction of BkO_2 or Bk_2O_3, but the quantity of berkelium oxide available for reduction at one time has not been large enough to produce other than thin foils by this technique. Such a form of product metal can be very difficult to handle in subsequent experimentation. The rate and yield of Am from the reduction at 1525 K of americium dioxide with lanthanum metal are given in Fig. 2.

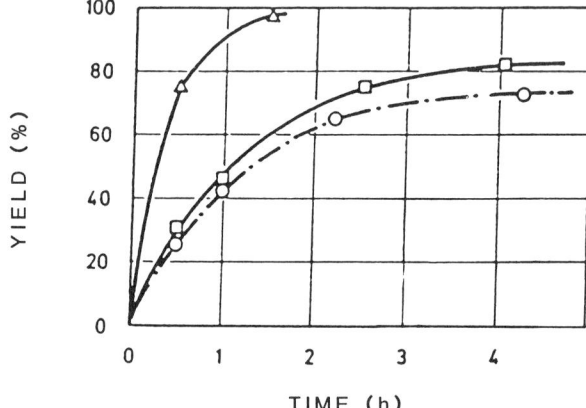

FIG. 2. Metallothermic reduction at 1525 K of AmO_2 by La metal as a function of time: ○ and □, mixtures of AmO_2 powder and La metal turnings; △, pelletized mixture of AmO_2 powder and ground La metal turnings.

Actinide metals with lower vapor pressures (Th, Pa, and U) cannot be obtained by this method since no reductant metal exists which has a sufficiently low vapor pressure and a sufficiently negative free energy of formation of its oxide. For the large-scale production of U, Np, and Pu metals, the calciothermic reduction of the actinide oxide (Section II,A) followed by electrorefining of the metal product is preferred (24). In this process the oxide powder and solid calcium metal are vigorously stirred in a $CaCl_2$ flux which dissolves the by-product CaO. Stirring is necessary to keep the reactants in intimate contact.

C. Metallothermic Reduction of Actinide Carbides

Transition metals like Ti, Zr, Nb, and Ta are able to reduce actinide carbides to metals according to the following generic reaction:

$$AnC + Ta \longrightarrow TaC + An$$

The free energies of formation of the transition metal carbides are somewhat more negative than the free energies of formation of the actinide carbides. To facilitate separation of the actinide metal from the reaction products and excess transition metal reductant, a transition metal with the lowest possible vapor pressure is chosen as the reductant. Tantalum metal and tantalum carbide have vapor pressures which are low enough (at the necessary reaction temperature) to avoid contamination of the actinide metal by co-evaporation.

Actinide carbides are prepared by carbothermic reduction of the corresponding dioxides according to the reaction:

$$AnO_2 + 3C \longrightarrow AnC + 2CO$$

In practice, a mixture of actinide dioxide and graphite powder is first pelletized and then heated to 2275 K in vacuum in a graphite crucible until a drop in the system pressure indicates the end of CO evolution. The resulting actinide carbide is then mixed with tantalum powder, and the mixture is pressed into pellets. The reduction occurs in a tantalum crucible under vacuum. At the reduction temperature, the actinide metal is vaporized and deposited on a tantalum or water-cooled copper condenser.

The yield and rate of the tantalothermic reduction of plutonium carbide at 1975 K are given in Fig. 3. Producing actinide metals by metallothermic reduction of their carbides has some interesting advantages. The process is applicable in principle to all of the actinide metals, without exception, and at an acceptable purity level, even if quite impure starting material (waste) is used. High decontamination factors result from the selectivities achieved at the different steps of the process. Volatile oxides and metals are eliminated by vaporization during the carboreduction. Lanthanides, Y, Ti, Zr, Hf, V, Nb, Ta, Mo, and W form stable carbides, whereas Rh, Os, Ir, Pt, and Pd remain as nonvolatile metals in the actinide carbides. Thus, these latter elements

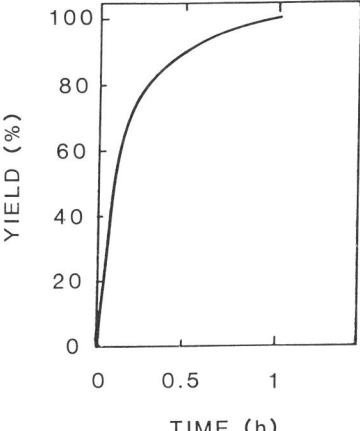

FIG. 3. Metallothermic reduction at 1975 K of PuC with Ta metal as a function of time.

$$PuC + Ta \rightarrow TaC + Pu(\uparrow)$$

will contaminate the actinide carbides if they are present in the starting materials. The carbides of Zr, Hf, Nb, and W are not reduced by Ta and remain as stable carbides in the reaction products during the tantalothermic reduction. The other elements are eliminated during selective vaporization and deposition of the actinide metal (Section III,B).

This process is particularly useful for the preparation of pure plutonium metal from impure oxide starting material (*111*). It should also be applicable to the preparation of Cm metal. Common impurities such as Fe, Ni, Co, and Si have vapor pressures similar to those of Pu and Cm metals and are difficult to eliminate during the metallothermic reduction of the oxides and vaporization of the metals. They are eliminated, however, as volatile metals during preparation of the actinide carbides.

The light actinide metals (Th, Pa, and U) have extremely low vapor pressures. Their preparation via the vapor phase of the metal requires temperatures as high as 2375 K for U and 2775 K for Th and Pa. Therefore, uranium is more commonly prepared by calciothermic reduction of the tetrafluoride or dioxide (Section II,A). Thorium and protactinium metals on the gram scale can be prepared and refined by the van Arkel–De Boer process, which is described next.

D. Iodide Transport (van Arkel–De Boer) Process

The van Arkel–De Boer process is widely used to refine metals. A transporting agent such as I_2 reacts with the metal (M) to be refined to form a volatile iodide. This iodide is then decomposed at a higher temperature into the refined metal and I_2, which becomes available again to react with the impure metal, thus sustaining the process:

$$M + 2I_2 \longrightarrow MI_4 \quad \text{(low temperature)}$$

$$MI_4 \longrightarrow M + 2I_2 \quad \text{(high temperature)}$$

For successful use of the van Arkel–De Boer process starting with an actinide compound, it is necessary that the original actinide compound react readily with I_2 to yield a volatile actinide iodide. Both ThC and PaC, easily prepared from the corresponding metal oxides by carboreduction (Section II,C), react with I_2 at 625 K to yield volatile iodides and carbon. Above 1475 K these iodides are unstable and decompose into the respective metals and iodine.

Expressed in chemical equations for Th,

$$ThC + 2I_2 \longrightarrow ThI_4 + C \quad (625 \text{ K})$$

$$ThI_4 \longrightarrow Th + 2I_2 \quad (\text{above } 1475 \text{ K})$$

Using Pyrex ampoules with resistively heated tungsten wire or strip filaments, protactinium metal has been prepared on the milligram scale (9, 13, 15). An improved technique is to use a quartz van Arkel–De Boer bulb with an inductively heated W sphere which solves the previous problem of filament breaking and considerably improves the deposition rate of Pa metal (109).

Proceeding from thorium to plutonium along the actinide series, the vapor pressure of the corresponding iodides decreases and the thermal stability of the iodides increases. The melting point of U metal is below 1475 K and for Np and Pu metals it is below 975 K. The thermal stabilities of the iodides of U, Np, and Pu below the melting points of the respective metals are too great to permit the preparation of these metals by the van Arkel–De Boer process.

E. MOLTEN SALT ELECTROLYSIS

Methods have been developed (75) to prepare actinide metals directly from actinide oxides or oxycompounds by electrolysis in molten salts (e.g., LiCl/KCl eutectic). Indeed, the purest U, Np, and Pu metals have been obtained (19, 24) by oxidation of the less pure metal into a molten salt and reduction to purer metal (electrorefining, Section III,D).

III. Purification (Refining) of Actinide Metals

A. VACUUM MELTING WITHOUT DISTILLATION OF THE PRODUCT METAL

The metal to be refined is melted in high vacuum. The melt is kept at the highest possible temperature at which the vapor pressure of the liquid metal is still acceptably low in order to facilitate rapid evaporation of the volatile impurities. Vacuum melting is used to eliminate traces of iodides and iodine in metals prepared by the van Arkel–De Boer process (Th and Pa). Th, U, Np, and Pu metals obtained by the metallothermic reduction of the halides (Section II,A) are vacuum melted to eliminate traces of the reductant metals (Li, Ca, or Ba) and of the reaction by-products (alkali metal or alkaline earth halide salts). Electrolyte inclusions in molten salt-electrorefined metals (U, Np, Pu) can be eliminated by vacuum melting. Melting the metal in ultrahigh vacuum is necessary due to the reactive nature of the actinide metals. Melting of actinides in W or Ta crucibles results in some contamination from the crucible material. In the case of Pu metal at 1000 K, the contamination is slight (less than 50 atomic ppm when W crucibles are used). It is less than 10 atomic ppm when Ta crucibles coated with tantalum oxide are used (60). Levitation melting techniques (86, 95) have been used to avoid contamination of the actinide metal by the crucible material (1).

B. SELECTIVE VAPORIZATION

Efficient refining of the more volatile actinide metals (Pu, Am, Cm, Bk, and Cf) is achieved by selective vaporization for those (Pu, Am, Cm) available in macro quantities. The metal is sublimed at the lowest possible temperature to avoid co-evaporation of the less volatile impurities and then deposited at the highest possible temperature to allow vaporization of the more volatile impurities. Deposition occurs below the melting point of the metal to avoid potential corrosion of the condenser by the liquid metal. Very good decontamination factors can be obtained for most metallic impurities. However, Ag, Ca, Be, Sn, Dy, and Ho are not separated from Am metal nor are Co, Fe, Cr, Ni, Si, Ge, Gd, Pr, Nd, Sc, Tb, and Lu from Cm and Pu metals.

Nonmetallic impurities, mostly oxygen, found in actinide metals distilled under a vacuum of 0.1 mPa range from 4000 to 7000 atomic ppm. In a vacuum of 0.1 μPa the nonmetallic impurity content decreases to between 400 and 880 atomic ppm (51, 52).

Selective vaporization and deposition are performed in radiofrequency-heated tantalum distillation columns (Fig. 4). The

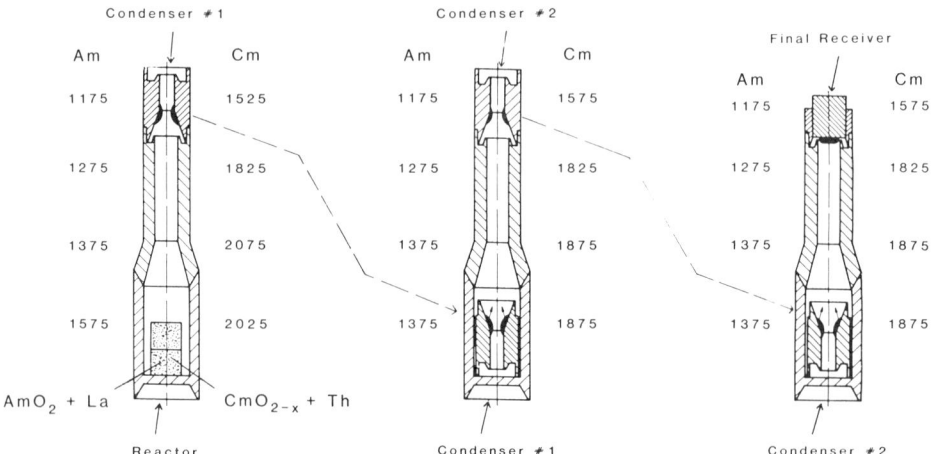

FIG. 4. Tantalum distillation columns and appropriate temperatures (K) for the preparation–purification of Am and Cm metals.

evaporation rate is maintained at the ideal value of 1–5 g of metal per hour. Americium is distilled at 1375 K and curium at 1875 K. The deposition temperatures are 1175 K for americium metal and 1575 K for curium metal (*112*).

C. IODIDE TRANSPORT

The efficiency of the van Arkel–De Boer process (Section II,D) for refining thorium and protactinium metals can be increased by repeating the process to achieve higher purity of product metal.

D. ELECTROREFINING, ZONE MELTING, AND SOLID-STATE ELECTROTRANSPORT

If an actinide metal is available in sufficient quantity to form a rod or an electrode, very efficient methods of purification are applicable: electrorefining, zone melting, and electrotransport. Thorium, uranium, neptunium, and plutonium metals have been refined by electrolysis in molten salts (*84*). An electrode of impure metal is dissolved anodically in a molten salt bath (e.g., in LiCl/KCl eutectic); the metal is deposited electrochemically on the cathode as a solid or a liquid (*19, 24*). To date, the purest Np and Pu metals have been produced by this technique.

In zone melting, a narrow molten zone is passed several times along a thin metal rod. Metallic impurities and carbon dissolve in the liquid and move with the molten zone to the end of the rod, whereas oxygen and nitrogen move to the opposite end.

If the metal is kept just below its melting temperature, refining in an electric field [solid-state electrotransport processing (SSEP)] is possible for actinide metals with low vapor pressures (Th, Pa, U, Np, and Pu). Oxygen, nitrogen, and carbon are carried with the electron flow, while metallic impurities move in the opposite direction, resulting in purification of the central part of the rod. Both zone melting and SSEP require a very high vacuum. Refining of Th by SSEP (1875 K, ~2 × 10^3 A/cm^2, 0.1 mPa) resulted in metal with a resistivity ratio (Sections I and V,B) of 200;[1] improving the vacuum to 0.3 nPa yielded Th metal with a resistivity ratio of about 2000 (*99*). A Th metal rod was then made from four segments of this best, once-processed Th and reprocessed by SSEP at 0.3 nPa to produce the world's purest Th metal, that with a resistivity ratio of 4200 (*99*).

---

[1] The *resistivity ratio* is defined as the resistivity of the metal at 300 K divided by its resistivity at 4.2 K. The resistivity at 4.2 K is assumed to be due entirely to impurities.

IV. Growth of Single Crystals of Actinide Metals

Actinide solid-state research progress depends upon the availability of single crystals of the highest possible purity and perfection level. Due to spin–orbit coupling, the properties of actinide compounds are very anisotropic. The unusually large anisotropic field (10^6 Oe or higher) cannot be explained by orbital anisotropy alone. Investigation of this phenomenon (bonding anisotropy) can only be carried out on good single crystals by magnetization measurements, neutron scattering, and angular- and energy-resolved photoemission spectroscopy.

Single crystals of the actinide metals are particularly difficult to grow. The light actinide metals (U, Np, and Pu) exhibit multiple crystallographic phases. Methods of crystal growth from the liquid state (zone melting, Czochralski) are therefore not usually applicable. More delicate crystal growth techniques (64) like recrystallization in the solid state using grain coarsening (39), strain annealing (16, 116), solid-state electrotransport (99), or phase transformation (49, 90, 105) techniques at normal or high pressure (70, 71) have to be employed. The various methods used to grow single crystals of actinide metals are summarized in Table IV. The van Arkel–De Boer process (chemical vapor transport) has facilitated the preparation of single crystals of thorium and protactinium metals (Fig. 5) (108, 109). The best single crystals of actinide metals available in macro quantities are obtained by solid-state electrotransport under ultrahigh vacuum (≤ 0.5 nPa).

TABLE IV

METHODS USED TO GROW SINGLE CRYSTALS OF ACTINIDE METALS

Element	Method	Reference
Thorium	Strain anneal	*116*
	Phase change	*49, 90*
	van Arkel–De Boer (chemical vapor transport)	*108*
	Solid-state electrotransport	*99*
Protactinium	van Arkel–De Boer (chemical vapor transport)	*109*
Uranium	Strain anneal	*16*
	Phase change	*105*
	Grain coarsening	*39*
	High pressure	*71*
	Electrodeposition in molten salts	*17*
Neptunium	—	None reported
Plutonium	High pressure	*70*
Americium	Physical vapor transport	*106*
Curium	Physical vapor transport	*83*

FIG. 5. Scanning electron micrograph of single crystals of Pa metal grown by iodide transport (Sections II,D and III,C); magnification = 100 × (*107*).

Large single crystals (54 mm^3) of ultrapure thorium metal have been prepared by this technique (*99*).

During the electrorefining of uranium metal in a molten salt eutectic, a low current density favors the formation of large single crystals. Up to 5-cm^3 single crystals of uranium metal have resulted from the large-scale (100 kg of U) electrorefining of uranium metal in molten LiCl/KCl eutectic (*17*).

Single crystals of elemental Am (*106*) and Cm (*83*) have been obtained during their refining by deposition of the metal vapor at 1175 K and 1575 K, respectively (Section III,B; Fig. 4).

V. Specifics of Actinide Metal Preparation

A. GENERAL COMMENTS

Differences in the availability and radioactivity of the actinides lead to a great disparity in the lengths of the following discussions of the individual elements. For instance, both thorium and uranium have been known since the nineteenth century; there seems little point in discussing their early histories, and we have chosen not to do so, which results in a rather brief section for each. The methods developed for neptunium metallurgy are so strongly based on the extensive developmental work carried out for the much more widely used element plutonium that an extensive discussion for neptunium would be redundant. In fact, the methods used for plutonium metallurgy are generally applicable to thorium, uranium, and neptunium, and need not be redescribed. At the heavy end of the actinides, the elements beyond einsteinium exist only in submicrogram or even in so little as few-atom quantities. Some fascinating techniques have been worked out to examine the nature of these elements, but we have chosen to exclude these studies as being beyond the scope of this review.

B. ACTINIUM

Although in a strict sense not a member of the actinide series, actinium is included here for completeness as lanthanum is frequently included in the lanthanide series. The longest lived isotope of actinium is ^{227}Ac with a half-life of 21.8 years. It occurs in nature from the decay of ^{235}U but also may be prepared by bombarding ^{226}Ra with neutrons. Actinium decays via a series of short-lived isotopes, eventually ending with stable lead. The presence of these radioactive daughters, particularly ^{227}Th (which is a strong γ-emitter), necessitates the use of lead-lined gloved boxes and remote control manipulators. Consequently, the metallurgy of actinium has been little studied and, due to the great expense and trouble involved, probably will not be studied extensively in the future.

The first preparation of metallic Ac was on the microgram scale and used the metallothermic reduction (Section II,A) of $AcCl_3$ with K metal vapor (38), which is the same method used by Klemm and Bommer (67) to prepare La metal. The metal produced by this method is mixed with KCl and K metal. X-Ray diffraction revealed that Ac metal was isostructural with β-La, but that the face-centered cubic cell dimension of Ac (5.311 Å) was slightly larger than that of La (5.304 Å).

A milligram-scale preparation has been reported that used AcF_3 with Li metal as the reductant (115). No X-ray diffraction data were obtained. The melting point was found to be 1325 ± 50 K. The metal was observed to glow in the dark with a blue color (115).

Reduction of Ac_2O_3 by Th metal followed by evaporation of the Ac metal (Section II,B) was used by Colson (9, 22); the product was re-evaporated and then collected on tantalum platelets. It was characterized by X-ray diffraction. Two different preparations gave values for the unit cell dimension of 5.317 and 5.314 Å, slightly larger than the value reported previously (38).

In principle, a promising method for the preparation of Ac metal is the tantalothermic reduction of AcC, as described generally in Section II,C. This method has not been tried as yet, however, so the metallothermic reduction of an actinium halide or oxide remains the only proved method.

C. Thorium

Thorium occurs naturally primarily as the very long-lived ^{232}Th ($\tau_{1/2} > 10^{10}$ years), along with small amounts of other isotopes resulting from the radioactive decay of ^{235}U and ^{238}U. It is so weakly radioactive that very few precautions are necessary in its handling.

Thorium metal is generally prepared by the metallothermic reduction of its halides (Section II,A). Very high-quality metal containing a total of 250 ppm impurities has been prepared at the Ames Laboratory of the Department of Energy (98, 99). These workers reduced $ThCl_4$ with excess Mg metal to yield a Th–Mg alloy, which was then heated *in vacuo* to remove the excess Mg (99):

$$ThCl_4 + Mg(xs) \longrightarrow Th\text{–}Mg(alloy) + 2MgCl_2$$

$$Th\text{–}Mg(alloy) \xrightarrow[\Delta]{vacuum} Th(sponge) + Mg(vapor)$$

Ultrapure Th metal has been processed at the Ames Laboratory by solid-state electrotransport under very low pressures (on the order of 0.3 nPa), which has produced the purest Th metal known, that with a resistivity ratio of 4200 for doubly refined metal (99–101). This resistivity ratio of 4200 translates into probably <50 ppm total impurities in the metal (see footnote 1) (87–90, 104). Single crystals measuring 0.25 cm in diameter by 1.1 cm in length with resistivity ratios of 1700–1800 have also been grown (99).

The method of choice for the preparation of Th metal is reduction of the tetrachloride (Section II,B) by Mg (99), followed by refinement using electrotransport purification (Section III,D) (87, 88, 90).

D. Protactinium

The most common isotope of protactinium is ^{231}Pa ($\tau_{1/2} = 3.3 \times 10^4$ years), which occurs in pitchblende in the amount of 300 mg/ton, about the same as radium. The heroic efforts of British researchers resulted in the isolation of some hundred grams of ^{231}Pa from the sludge left over from uranium processing; without this supply, little or nothing would be known about Pa metal. Freshly purified Pa can be handled in gloved boxes without extra shielding. Aged Pa emits strong and intense penetrating radiation resulting from its decay products so that shielding must be provided.

Protactinium metal was first prepared in 1934 by thermal decomposition of a pentahalide on a hot filament (50). It has since been prepared from PaF_4 by metallothermic reduction (Section II,A) with barium (26, 27, 34, 102), lithium (40), and calcium (73, 74). However, the highest purity metal is achieved using the iodide transport (van Arkel–De Boer) process (Section II,D).

The method of choice for the preparation of Pa metal is a somewhat modified van Arkel–De Boer process, which uses protactinium carbide (Section II,C) as the starting material. The carbide and iodine are heated to form protactinium iodide, which is thermally dissociated on a hot filament (12–15). An elegant variation is to replace the filament with an inductively heated W or Pa sphere (109). A photograph of a 1.4-g sample of Pa metal deposited on a radiofrequency-heated W sphere is shown in Fig. 6. From the analytical data presented in Table V, the impurities present before and after application of this modified iodide transport process (Sections II,D and III,C) can be compared.

Levitation melting (Section III,A) of Pa metal in high vacuum results in a considerable increase in purity. Metallic Pa resembles Th metal in that it has a very high melting point and a low vapor pressure.

FIG. 6. A 1.4-g mass of Pa metal crystals deposited on a W sphere.

TABLE V

VAN ARKEL–DE BOER PURIFICATION OF PROTACTINIUM[a]

Element	Elemental impurities (atomic ppm)[b]	
	Starting material (oxide)	Refined metal
B	600	<1
Mg	600	25
Al	1,400	55
Si	68,000	600
Fe	2,500	100
Cu	700	6
Sr	100	<1
Nb	1,800	<1
Mo	180	5
Ag	150	1
Ba	1,600	<2
Bi	48	<1
Th	120	97
U	74	29
Am	60	<1
Ti	—	170
Total purity	93 at %	99.7 at %

[a] From ref. (*109*).
[b] All metals not listed were <100 atomic ppm. Nonmetallic elements in the refined metal were found to be 200 atomic ppm for N and 300 atomic ppm for O, with only traces of other elements.

Secondary refining processes such as zone melting and solid-state electrotransport (Section III,D) should yield ultrahigh-purity Pa metal.

E. URANIUM

Uranium is spread widely over the world but usually occurs in very low concentrations. Since the beginning of the nuclear age, the estimates of its abundance have been revised upwardly rather dramatically. Presently there is a surfeit, which is probably a temporary situation. In any event, multiton amounts are available, which affects the technology employed in its metallurgy. The radioactivity of natural uranium is so low as to present no problems in its handling. Over 99% is composed of very long-lived ^{238}U with a half-life of 4.5×10^9 years; the balance is primarily the fissionable isotope ^{235}U. The real concern associated with uranium handling probably stems more from the danger of heavy metal poisoning than from radioactivity.

It appears that Peligot in 1841 was the first to prepare U metal, using the metallothermic reduction of UCl_4 by K metal (Section II,A). At present, U metal is usually prepared by the same general method; that is, reduction of uranium halides by alkali or alkaline earth metals. The calciothermic reduction of UF_4 in either closed bombs or open crucibles is the most commonly used method for preparing U metal (121). The product metal is usually relatively impure. When special precautions were taken to purify the reagents, multigram quantities of very pure U metal containing only a few hundred ppm of impurities (Table VI) were obtained by bomb reduction of UF_4 with Ca metal (7, 65). Further refining can improve the quality of the metal. Vacuum melting (Section III,A) is used to eliminate the excess reductant metal. Electrorefining (Section III,D) in molten salts such as LiCl/KCl eutectic (11) is also a very effective purification method. At low current densities and on the kilogram scale, large single crystals (Section IV) of pure U metal have been obtained by this secondary purification (17).

As with Th and Pa metals, ultrapure U metal can in principal be obtained by such methods as zone melting and solid-state electro-

TABLE VI

AVERAGE ANALYSIS OF URANIUM METAL PRODUCED BY Ca REDUCTION OF UF_4 ON THE 250-g SCALE[a]

Element	Impurity level (ppm by wt)
Al	<2
B	<0.1
Be	<0.1
C	<25
Ca	<10
Co	<5
Cr	2
Cu	1
Fe	50
Li	<0.1
Mg	<3
Mn	6
Na	<1
Ni	8
O	<70
Si	<7
V	<10
Total	<200

[a] From ref. (65).

transport (Section III,D). The method of choice for the preparation of excellent-quality U metal is bomb reduction of UF$_4$ by Ca metal (7, 65).

F. NEPTUNIUM

The only isotope of Np suitable for chemical work is ^{237}Np, which has a very long $\tau_{1/2}$ of 2.14×10^6 years. It is an α-emitter with some penetrating γ radiation, and is available in kilogram amounts. The isotope is formed in nuclear reactors from both ^{235}U and ^{238}U and also results from the α-decay of ^{241}Am. Although it is the least toxic of the common transuranic isotopes, it is about 1000 times more radioactive than U and should always be handled in gloved boxes.

The metallothermic reduction of NpF$_3$ with Ba metal vapor (Section II,A) was used in 1948 to prepare Np metal on the 50-μg scale (41). The same method was used to prepare some few milligrams of elemental Np in 1951 (124). The metal has been prepared on the gram and multigram scales with yields up to 99% by reduction of NpF$_4$ with Ca metal (4, 33, 35, 68, 76); Li metal has also been used as the reductant (78). The metal obtained by these methods is not very pure, but it can be refined further by vacuum melting (Section III,A) and by electrorefining in molten salts (Section III,D). Another preparative method is electrolysis between a graphite anode and a W cathode of Np(III) dissolved in a molten LiCl/KCl eutectic (Section II,E) (75). Figure 7 shows 2.5 g of Np metal electrodeposited on a W electrode.

Direct oxide reduction by Ca metal (Section II,A) in a molten CaCl$_2$ solvent system (80) has been used for kilogram-scale production of Np metal. The product metal is further purified by electrorefining (Section III,D). This combination has been used to prepare Np metal which is 99.9 at % pure. Analyses of two preparations are given in Table VII (20). A

FIG. 7. Neptunium metal (2.5 g) electrodeposited on a W electrode.

TABLE VII

ANALYSIS OF ULTRAPURE NEPTUNIUM METAL[a]

Element	Impurity level (ppm by wt)	
	Preparation 1	Preparation 2
Ag	3	5
Al	30	25
Am	0.2[b]	0.2[b]
B	<5	<5
Be	<1	<1
Bi	<1	<1
C	20	5
Ca	100	300
Cd	<10	<10
Cr	<5	<5
Cu	<1	<1
Fe	<5	<5
Ir	<100	<100
Mg	<1	<1
Mn	<1	<1
Mo	<20	<20
Na	<50	<50
Ni	<5	<5
O	75	40
Pb	<5	<5
Pu	14	14.6
Si	<5	<5
Sn	<5	<5
Ta	<10	<10
U	<10	<10
W	<10	<20
Zn	<5	<5
Total purity (wt %)	99.97	99.96

[a] From ref. (20).
[b] On December 10, 1984.

photograph of a 500-g button of Np metal produced in this manner is shown in Fig. 8.

The method of choice for kilogram-scale preparations of Np metal is direct oxide reduction by Ca metal in a molten $CaCl_2$ solvent system as described above, followed by electrorefining. This metal can be further refined by levitation melting (Section III,A).

Unfortunately, a transition from the α phase of Np metal to the β phase occurs at the low temperature of 301 K, making the growth of single crystals of α-Np metal extremely difficult.

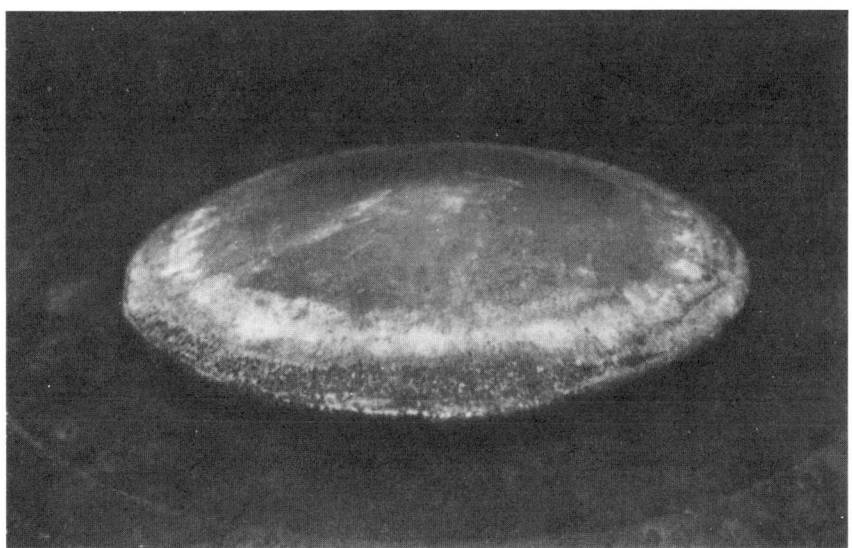

FIG. 8. A 500-g button of Np metal. Courtesy Los Alamos National Laboratory, Los Alamos, New Mexico.

G. PLUTONIUM

The relatively long-lived isotopes of Pu suitable for chemistry and metallurgy are those of masses 238, 239, 240, 241, 242, and 244. Plutonium formed in nuclear reactors occurs as a mixture of isotopes. A typical isotopic composition of Pu in spent fuel containing 10.4 kg of Pu/ton of fuel is given in Table VIII.

TABLE VIII

TYPICAL ISOTOPIC COMPOSITION OF PLUTONIUM IN SPENT REACTOR FUEL[a]

Mass number	Weight percent	$\tau_{1/2}$ (years)	Principal radiation
238	1	87.7	α
239	56	2.4×10^4	α
240	26	6.6×10^3	α
241	12	14.4	β
242	5	3.8×10^5	α

[a] From ref. (18).

The isotope ^{238}Pu is of special value as a heat source to make electricity for various esoteric uses, mainly in space. Plutonium-244, with a $\tau_{1/2}$ of 8.1×10^7 years, is the longest lived Pu isotope available in multigram quantities. It is formed from various Pu isotopes by successive neutron capture. Unfortunately, it is very rare and expensive due to the fact that long irradiations are necessary to produce significant amounts and that it has to be separated from the mixture of Pu isotopes formed in nuclear reactors. Long-lived ^{242}Pu can be prepared by extremely long-term neutron irradiation of reactor Pu so that the isotopes of mass numbers 239–241 are burned up or converted to ^{242}Pu. This isotope can be made 99% isotopically pure by neutron irradiation of ^{241}Am to produce ^{242}Am, which has a 17% electron capture decay branch to ^{242}Pu. It is available in multigram quantities and thus is extremely valuable, although very expensive, for research purposes.

It should be emphasized that all isotopes of Pu are sufficiently radioactive so that they must be handled in gloved boxes. However, Pu is formed as a mixture of isotopes, many of which emit neutrons and/or form daughters that emit penetrating radiation resulting from their decay. Neutron emission rates of various Pu isotopes are given in Table IX (103).

Because of the wide variations in the neutron emission rates shown in Table IX, it is obviously necessary to know the isotopic composition of any particular batch of Pu in order to know how much and what kind of shielding is demanded for biologic safety.

The first Pu metal ever made was prepared in late November 1943 by metallothermic reduction of about 35 μg of PuF$_4$ with Ba metal (Section II,A) (42). This method, modified to use Ca as the reductant for PuF$_4$, has been widely used and is very successful; yields of more than 98% and purities of the order of 99.8 at % are obtained (2, 5, 62, 81, 96). The Pu

TABLE IX

Neutron Emission Rates for Various Plutonium Isotopes[a]

Isotope	From spontaneous fission ($n\ \sec^{-1}\ g^{-1}$)	From (α,n) reactions in PuF$_4$ ($n\ \sec^{-1}\ g^{-1}$)
^{238}Pu	3.4×10^3	2.1×10^6
^{239}Pu	0.02	4.3×10^3
^{240}Pu	1.0×10^3	1.6×10^3
^{242}Pu	1.7×10^3	1.7×10^2

[a] From ref. (103).

metal from the calciothermic reduction can be refined by vacuum melting (Section III,A) or by electrorefining in molten salts (Section III,D) to give metal of 99.9+ at % purity.

Another approach used to prepare massive amounts of Pu metal is reduction of PuO_2 with Ca metal (Section II,A) in a molten $CaCl_2$ flux. The molten $CaCl_2$ dissolves the CaO and separates the metallic phases (118). The yield and purity of the product Pu metal are excellent. This method was further developed and then used to prepare massive amounts of elemental ^{238}Pu for isotopic power sources (84). This DOR method (6, 19, 21, 24, 84) is now competitive with the metallothermic reduction of PuF_4 described above. A photograph of a typical Pu metal product from this DOR process is shown in Fig. 9. Electrorefining of this metal routinely yields Pu of 99.9 at % purity. A photograph of a cylinder of ultrapure Pu metal produced by electrorefining is given in Fig. 10. An analysis of ultrapure Pu metal (99.995% by weight) furnished to the United States Bureau of Standards as a primary standard is given in Table X (60).

The method of choice for the preparation of Pu metal on the multigram to kilogram scale is metallothermic reduction of PuF_4 or PuO_2 by Ca metal (69), followed by electrorefining with TaC electrodes and vacuum casting. This method produces metal 99.9 at % pure. An

FIG. 9. Photograph of a typical Pu metal product from the DOR process.

FIG. 10. 5.3 kg of ultrapure (99.99% by wt) Pu metal produced by electrorefining in a molten salt.

alternative method is the tantalothermic reduction of PuC (*111*) followed by selective vaporization and deposition (Section III,B). A review of preparative methods for Pu metal has been published (*69*).

Plutonium metal exhibits the most complex series of crystalline phases of any known element. At normal pressure there are six metallic phases of Pu between room temperature and its 913 K melting point (*60*).

Single crystals of α-Pu have been grown by holding polycrystalline Pu metal at 675 K under a hydrostatic pressure of 5.5 GPa for several days (*70*).

H. AMERICIUM

The isotopes of Am useful for metallurgical studies are ^{241}Am and ^{243}Am, both of which are available in multigram or even kilogram amounts. The lighter isotope, ^{241}Am, has the shorter $\tau_{1/2}$ of only 432 years, which results in self-heating and consequent corrosion of the metal. It is thus less desirable for metallurgical studies. This isotope is formed by β-decay of ^{241}Pu ($\tau_{1/2} = 14$ years), which itself is formed by

TABLE X

IMPURITY LEVELS IN PLUTONIUM[a,b]

Element	Concentration[c]	Element	Concentration[c]
Li	<0.005	Cd	<1
Be	<0.001	Sn	<1
B	<0.5	Ba	<0.5
Na	<1	La	<1
Mg	2	Hf	<1
Al	<0.5	Re	<1
K	<0.5	Pb	<1
Ca	<4	Bi	<1
Ti	<0.5	Fe	5
V	<0.5	Am	7
Cr	<0.5	Si	2
Mn	<0.5	C	10
Co	<1	U	11
Ni	<0.5	Ta	6
Cu	<0.5	Th	<0.5
Zn	<5	Ga	<1
Sr	<0.5	W	<2.5
Y	<1	O	10
Zr	<1	H	<5
Mo	<1	N	3

[a] Casting made for the United States Bureau of Standards.
[b] From ref. (60).
[c] Expressed in parts per million of plutonium by weight.

successive neutron captures in a nuclear reactor. The second, heavier isotope, ^{243}Am, has a $\tau_{1/2}$ of 7380 years, 17 times longer than that of ^{241}Am, a fact which makes it much more suitable for metallurgical studies. This heavier isotope is formed by successive neutron captures in a nuclear reactor; its immediate precursor is 16-hour ^{242}Am. Both isotopes emit α- and γ-radiation and must be handled in shielded gloved boxes.

Metallothermic reduction of AmF_3 with elemental Ba (Section II,A) was used for the early, mostly microgram scale, preparations of Am metal (48, 72, 122, 123). Although the Am was not of high purity, the studies of these early Am metal products yielded valuable information about the basic thermodynamic, magnetic, and crystallographic properties of Am metal. In 1960, the first milligram quantities of Am metal were prepared using this method (79). At that time, workers at Los Alamos developed a new preparative method (79) based on the greater volatility of Am metal versus that of elemental La. Reduction of AmO_2

with La metal at elevated temperature resulted in the preparation and separation of the more volatile Am metal, which was then deposited on a fused silica fiber (Section II,B). The less-volatile by-product, La_2O_3, and excess La metal reductant were left behind in the Ta crucible. The size of the Am metal unit cell was then determined by X-ray diffraction of the metal deposited on the fiber.

Much effort was expended in improving techniques for both the metallothermic reduction of the halide (77), usually with Ba metal, and by reduction of the oxide with La metal (63, 77, 110, 117). Two significant improvements have been reported (110), which improved both yield and purity. The first was to pelletize intimate mixtures of La metal and AmO_2 powders (Section II,B; Fig. 2). The second was to eliminate the use of quartz for collection of Am vapor and ceramic crucibles for melting, since both quartz and ceramics were shown to introduce several thousand ppm oxygen. Use of Ta for these operations yielded Am metal containing <250 ppm by weight of oxygen.

Other methods for the preparation of elemental Am, mostly variations on the above two, have been studied (1, 23, 63). Thermal decomposition of the intermetallic compound Pt_5Am has also been used to prepare Am metal (36, 82, 110).

At the present time, when gram to kilogram amounts of either Am isotope are available, the method of choice for the preparation of Am metal is the metallothermic reduction of AmO_2 with La (or Th) using a pressed pellet of the oxide and the reductant metal. An oxide reduction–metal distillation still system is shown schematically in Fig. 11. Yields of Am metal are typically >90% and purity levels equal or exceed 99.5 at %. Further purification of the product Am metal can be achieved by repeated sublimations under high vacuum in a Ta apparatus (Section III,B; Fig. 4). A photograph of 2 g of Am metal distilled in a Ta apparatus is given in Fig. 12.

I. CURIUM

The isotopes of Cm available in multigram quantities are ^{242}Cm with a $\tau_{1/2}$ of only 163 days and ^{244}Cm with a longer $\tau_{1/2}$ of 18.1 years, still inconveniently short for chemical studies. These isotopes are made by successive neutron captures of ^{239}Pu in a nuclear reactor. The two relatively long-lived isotopes of Cm, available only in few-mg amounts, are ^{246}Cm with a $\tau_{1/2}$ of 4730 years and ^{248}Cm with a $\tau_{1/2}$ of 3.4×10^5 years. The cost of these two isotopes, measured in terms of neutrons used in their formation, is so great that it is probable that only milligram quantities will ever be available for study. The isotope ^{248}Cm is obtained in >90% isotopic purity from the α-decay of ^{252}Cf. All of

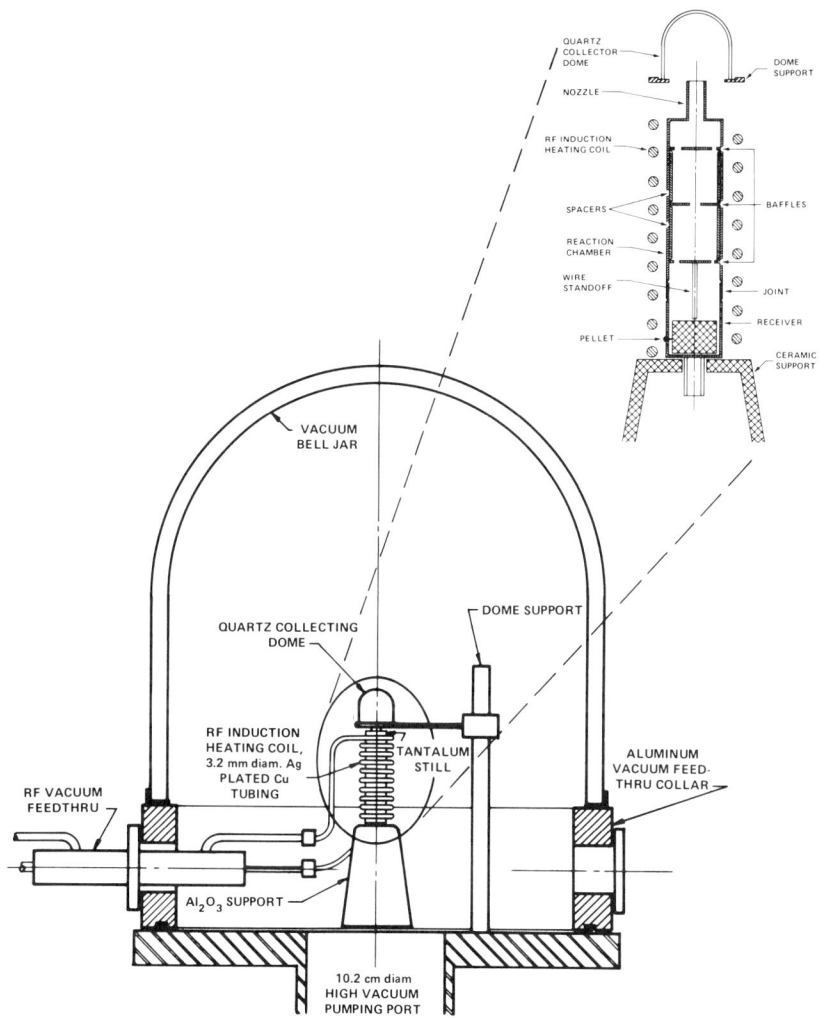

FIG. 11. An oxide reduction–metal distillation still system.

these curium isotopes are primarily α-emitters; however, the neutron emission rate accompanying spontaneous fission becomes more significant the larger the quantity of curium involved and/or the higher the mass number of the curium isotope. Also, the (α,n) reaction on the flourine nuclei in CmF_3 or CmF_4 produces neutrons which can be neglected with submilligram quantities of curium but which cause a serious problem with grams or more of curium.

Two preparations of Cm metal by the metallothermic reduction of CmF_3 with Ba metal (Section II,A) have been reported (*29, 119*). The

FIG. 12. Distilled Am metal (2 g) (center, front) and associated parts of the Ta still.

first, in 1951, resulted in the preparation of a few micrograms of a silvery metal which rapidly tarnished, presumably due to the high radioactivity of the ^{242}Cm employed. Thirteen years later, a similar reduction scheme using a double Ta crucible with an effusion hole resulted in the preparation of 0.25–0.50 mg of silvery ^{244}Cm metal. The contaminant found in the greatest concentration in this metal product was oxygen, at 4000 ppm.

To avoid the problems arising from excessive neutron production via (α,n) reactions when using a fluoride of Cm, a scheme employing reduction of the oxide by an Mg–Zn alloy in a flux of $MgCl_2$–MgF_2 was developed to produce up to gram quantities of elemental Cm (37). This method has the distinct disadvantage of concentrating any nonvolatile impurity originally present in the reagents in the product Cm metal.

Two other methods have been used successfully to prepare very pure Cm metal. A rather unique one is thermal decomposition of the intermetallic compound Pt_5Cm produced by hydrogen reduction of curium oxide in the presence of Pt (36, 82). The second method, the method of choice for gram-scale preparations of very pure Cm metal, involves reduction of curium oxide with Th metal (8, 83) in an apparatus like those shown schematically in Fig. 4 (Section III,B) and Fig. 11 (Section V,H). An intimate mixture of excess thorium metal and curium oxide is pelletized, outgassed, and then heated initially to 1925 K and finally to 2300 K (8). Yields of Cm metal of 74% (8) and about 90% (83) have been obtained. Extensions of this method, including multiple distillations and depositions at controlled temperatures (Section III,B; Fig. 4), resulted in the preparation of very high-purity Cm metal, with total cationic impurities at <50 ppm, O at 250 ppm, N at 20 ppm, and H at 10 ppm (83). A photograph of the Cm metal obtained by this method is shown in Fig. 13.

The method of choice for preparing up to 10 mg of metal, when using the rare ^{248}Cm, is metallothermic reduction (Section II,A) of anhydrous, oxygen-free CmF_4 (from treatment of CmF_3 with F_2 or ClF_3) with Ba metal vapor (31, 58, 113, 114).

J. BERKELIUM

The only isotope of Bk available in weighable quantities is ^{249}Bk, which is obtained by long-term neutron irradiation of Pu, Am, or Cm in

FIG. 13. A mass of Cm metal produced by distillation and deposition in a Ta still.

high-flux nuclear reactors. This isotope is a weak β-emitter, 126 keV maximum, which makes it difficult to identify when in the presence of other radioactivities, but which means that it poses little biologic danger until it decays (with a $\tau_{1/2}$ of 320 days) to the α-emitting ^{249}Cf. This decay daughter is so biologically dangerous that ^{249}Bk must be handled in gloved boxes.

Metallothermic reduction of BkF_3 by Li metal vapor (Section II,A) was used in 1968 for the first preparation of Bk metal on the few-microgram scale (94). This difficult task was accomplished by developing some unique microchemical techniques. Microgram quantities of Bk(III) were collected and manipulated by employing the single-bead ion-exchange resin technique (25). A W wire spiral holding the spherically shaped sample of BkF_3 was suspended from a Ta "chair," and it and the Li reductant were enclosed in a Ta crucible fitted with a cap having an effusion hole to permit the escape of by-product LiF and excess Li (92, 93).

The next preparation of Bk metal was in 1975 (43) and was on the much larger scale of 0.25–0.50 mg. The much improved yields resulted by lowering the heat capacity of the reduction system (113) to minimize vaporization losses of the relatively volatile Bk metal. This reduction system is shown schematically in Fig. 14. The most recent preparations of Bk metal (44, 53, 59) have been on the multimilligram scale and have used BkF_4 (instead of BkF_3) prepared by F_2 or ClF_3 treatment of BkO_2 or BkF_3. This method has the advantage of completely removing oxygen.

In principle, the DOR process using La or Th metal as the reductant (Section II,B) should be an excellent method for preparing Bk metal. However, the limited amount of Bk available (<50 mg/year) precludes

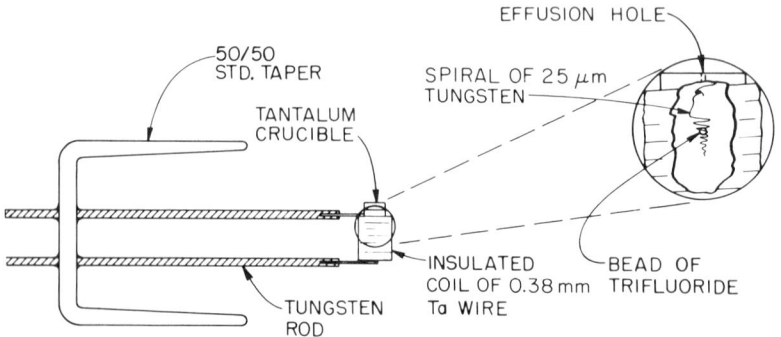

FIG. 14. Apparatus of low heat capacity used for preparation of Bk metal.

K. CALIFORNIUM

The only isotope of Cf suitable for chemical study is ^{249}Cf, an α-emitter with a $\tau_{1/2}$ of 351 years, which results from the β-decay of ^{249}Bk ($\tau_{1/2}$ = 320 days). This Cf isotope is an α-emitter and requires the usual precautions, such as working in gloved boxes, to avoid ingestion.

Initial attempts to prepare Cf metal using metallothermic reduction methods (Section II,A) were less than successful due to the high vapor pressure of Cf metal (28, 46). Reduction of californium oxide with La metal (Section II,B) and collection of the product Cf metal on a fused silica fiber (in the apparatus shown schematically in Fig. 15), were found to give metal with usable X-ray diffraction patterns (3). Later, the same method was used to collect Cf metal both on a fused silica fiber for X-ray diffraction analysis and on an electron microscopy grid for electron diffraction analysis (56). As more ^{249}Cf became available, preparations via this method were carried out on 0.4–1.0-mg samples of californium oxide (55), using fibers of quartz, Be, or C (suitable for direct X-ray diffraction analysis) to collect the product Cf metal.

A thorough study of the metallothermic reduction of CfF$_3$ with Li metal vapor (Section II,A) was reported in 1976 (85). From these studies, the three crystal structures of Cf metal and the temperature relationships between them were elucidated, but uncertainties still remained due to the lack of adequate analytical data concerning the level of impurities in the Cf metal studied (85).

All subsequent preparations of Cf metal have used the method of choice, that is, reduction of californium oxide by La metal and deposition of the vaporized Cf metal (Section II,B) on a Ta collector (10, 30, 32, 45, 91, 97, 120). The apparatus used in this work is pictured schematically in Fig. 16. Complete analysis of Cf metal for cationic and anionic impurities has not been obtained due to the small (milligram) scale of the metal preparations to date. Since Cf is the element of highest atomic number available for measurement of its bulk properties in the metallic state, accurate measurement of its physical properties is important for predicting those of the still heavier actinides. Therefore, further studies of the metallic state of californium are necessary.

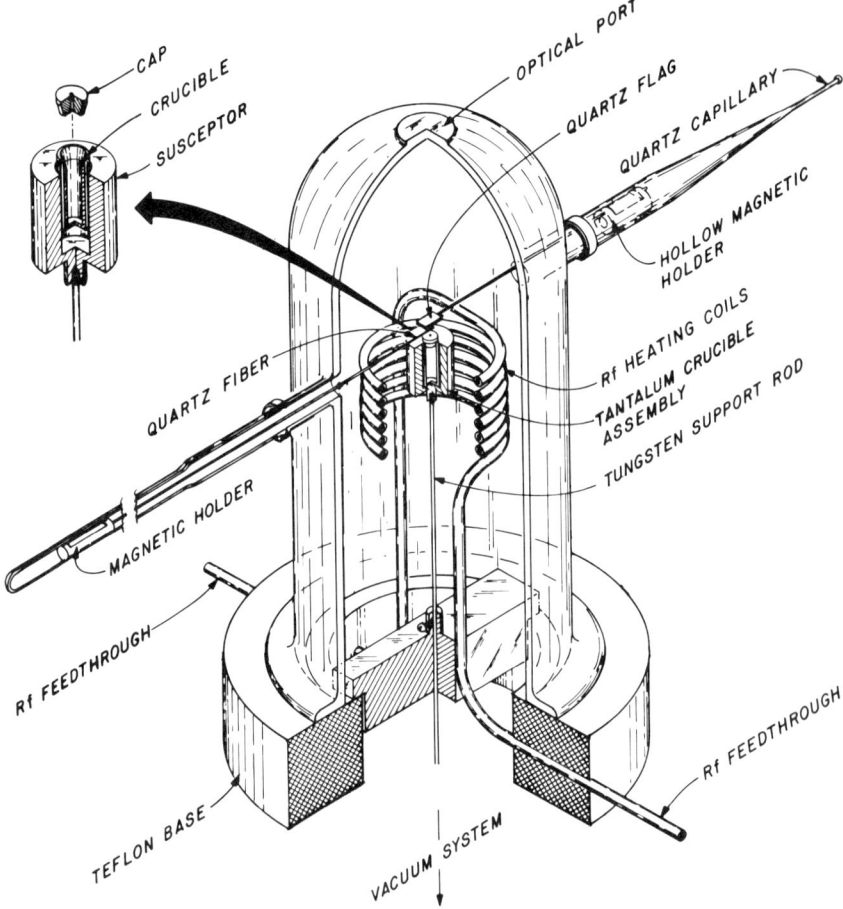

FIG. 15. Apparatus used for preparing early samples of californium metal.

L. Einsteinium

The only isotope of Es available in at least multimicrogram quantities is ^{253}Es, whose half-life is 20.5 days. Considering the 6.6-MeV α-particles it emits at the rate of about 6×10^{10} min^{-1} μg^{-1}, the α-decay energy alone amounts to some 1.5×10^4 kJ mol^{-1} min^{-1}. This explains why there have been very few attempts to prepare Es metal!

The first attempted preparation of elemental Es was made on the 1-μg scale by distilling Li metal *in vacuo* onto an unheated sample of EsF$_3$ (Section II,A), followed by quickly raising the temperature of the Li-

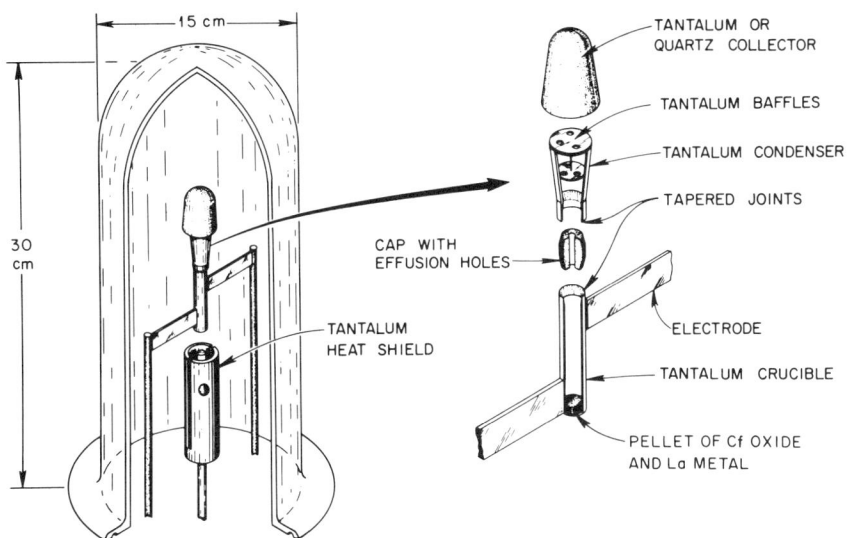

FIG. 16. Cf metal preparation apparatus.

coated EsF_3 to about 1075 K to promote reduction (28). The product obtained by quenching (to prevent complete loss of the volatile Es metal) exhibited four diffraction lines consistent with a face-centered cubic structure. These workers noted the high volatility of Es metal and suggested that it be exploited as a basis for quickly and efficiently separating and purifying Es from the lighter and less-volatile actinide metals. With the later availability of multimicrogram quantities of ^{253}Es, 20-μg pieces of Es_2O_3 were reduced with La metal (Section II,B) at about 1325 K; the evaporated metal was collected on carbon-coated, 3-mm diameter, electron microscopy grids (57). The thin film deposits were then analyzed by electron diffraction. The diffraction data were consistent with a face-centered cubic cell with a lattice parameter of 5.75 Å. Multimicrogram deposits of Es metal on Pt and Ta substrates have been obtained at Oak Ridge National Laboratory by distillation from Es_2O_3–La mixtures (54). The intense radiation and limited quantity of Es have precluded to date the preparation of bulk Es metal free of any supporting substrate.

The high volatility of metallic Es makes it an ideal candidate for preparation by metallothermic reduction of its oxide (Section II,B), but the scarcity of ^{253}Es prohibits its preparation in pure bulk form. Vaporization thermodynamics of Es metal have been determined, assuming that Henry's Law applies, using alloys of Es with divalent lanthanide metals (66). See Fig. 1 (Section II,A) for a comparison of the vapor pressures of many of the actinide metals, including Es.

VI. Some Physical Properties of Actinide Metals

Table XI gives the room-temperature, atmospheric pressure crystal structures, densities, and atomic volumes, along with the melting points and standard enthalpies of vaporization (cohesive energies), for the actinide metals. These particular physical properties have been chosen as those of concern to the preparative chemist who wishes to prepare an actinide metal and then characterize it via X-ray powder diffraction. The numerical values have been selected from the literature by the authors.

TABLE XI

Physical Properties of Actinide Metals

Actinide metal	Crystal structure	Density (g/cm^3)	Atomic volume (Å^3)	Melting point (K)	Enthalpy of vaporization ΔH°_{298} (kJ/mol)
Ac	Face-centered cubic	10.1	37.4	1325	(420)^a
Th	Face-centered cubic	11.72	32.85	2028	597
Pa	Body-centered tetragonal	15.42	24.89	1840	570
U	Orthorhombic	19.05	20.75	1405	531
Np	Orthorhombic	20.48	19.22	913	465
Pu	Monoclinic	19.86	20.00	913	343
Am	Double-hexagonal close packed	13.67	29.27	1448	284
Cm	Double-hexagonal close packed	13.51	29.98	1620	387
Bk	Double-hexagonal close packed	14.78	27.96	1260	310
Cf	Double-hexagonal close packed	15.1	27.4	1173	196
Es	Face-centered cubic	8.84	47.5	1133	131

^{a} Estimated value only; no measured value available.

Acknowledgments

The authors wish to thank the following colleagues for their cooperation and assistance in preparing this chapter: J. Reavis, J. Ward, C. Herrick, D. Christensen, and L. Mullins of the Los Alamos National Laboratory, Los Alamos, New Mexico; F. Schmidt and J. Smith of the Ames Laboratory, Iowa State University, Ames, Iowa; R. Haire of the Oak Ridge National Laboratory, Oak Ridge, Tennessee; and J. Fuger of the European Institute for Transuranium Elements, Karlsruhe, FRG.

This research has been sponsored in part by the Commission of the European Communities and in part by the Division of Chemical Sciences, United States Department of Energy under contracts DE-AS05-76ER04447 with the University of Tennessee (Knoxville), DE-AC05-84OR21400 with Martin Marietta Energy Systems, Inc., and W-7405-ENG-36 with the University of California.

References

1. Adair, H. L., *J. Inorg. Nucl. Chem.* **32**, 1173 (1970).
2. Anselin, F., in "Extractive and Physical Metallurgy of Plutonium and its Alloys" (W. D. Wilkinson, ed.), p. 61. Wiley (Interscience), New York, 1960.
3. Asprey, L. B., paper presented at the Third International Transplutonium Element Symposium, Argonne National Laboratory, Argonne, IL, October 20–22, 1971.
4. Baaso, D. L., Conner, W. V., and Burton, D. A., Doc. No. RFP-1032, Dow Chemical Co., Rocky Flats Div., U.S. Atomic Energy Commission, 1967.
5. Baker, R. D., and Maraman, W. J., in "Extractive and Physical Metallurgy of Plutonium and its Alloys" (W. D. Wilkinson, ed.), p. 43. Wiley (Interscience), New York, 1960.
6. Baldwin, C. E., and Navratil, J. D., *ACS Symp. Ser.* **216**, 369 (1983).
7. Bard, R. J., *et al.*, Doc. No. LA-1652, University of California, Los Alamos Scientific Laboratory, U.S. Atomic Energy Commission, Los Alamos, NM, 1954.
8. Baybarz, R. D., and Adair, H. L., *J. Inorg. Nucl. Chem.* **34**, 3127 (1972).
9. Baybarz, R. D., Bohet, J., Buijs, K., Colson, L., Müller, W., Reul, J., Spirlet, J. C., and Toussaint, J. C., *Transplutonium 1975, Proc. Int. Transplutonium Elem. Symp. 4th, 1975*, p. 61 (1976).
10. Benedict, U., Peterson, J. R., Haire, R. G., and Dufour, C., *J. Phys. F* **14**, L43 (1984).
11. Blumenthal, B., and Noland, R. A., *Prog. Nucl. Energy Ser. 5* **1**, 62 (1956).
12. Bohet, J., Rep. EUR 5882 FR, Commission of the European Communities, 1977.
13. Bohet, J., and Müller, W., *J. Less-Common Met.* **57**, 185 (1978).
14. Brown, D., Tso, T. C., and Whittaker, B., Rep. No. AERE-R 8638, United Kingdom Atomic Energy Research Establishment, 1976.
15. Brown, D., Tso, T. C., and Whittaker, B., *J. Chem. Soc., Dalton Trans.*, p. 2291 (1977).
16. Cahn, R. W., *Acta Metall.* **1**, 176 (1953).
17. Chauvin, G., Coriou, H., and Hure, J., *Met. Corros.-Ind.* **37**, 112 (1962).
18. Choppin, G. R., and Rydberg, J., in "Nuclear Chemistry, Theory and Applications," p. 511. Pergamon, New York, 1980.
19. Christensen, D. C., and Mullins, L. J., *ACS Symp. Ser.* **216**, 409 (1983).
20. Christensen, D. C., unpublished results, University of California, Los Alamos National Laboratory, Los Alamos, NM, 1985.
21. Christensen, E. L., Grey, L. W., Navratil, J. D., and Schulz, W. W., *ACS Symp. Ser.* **216**, 349 (1983).

22. Colson, L., Ph.D. Thesis, University of Liège, Belgium, 1975.
23. Conner, W. V., Doc. No. RFP-1188, Dow Chemical Co., Rocky Flats Div., U.S. Atomic Energy Commission, 1968.
24. Coops, M. S., Knighton, J. B., and Mullins, L. J., *ACS Symp. Ser.* **216**, 381 (1983).
25. Cunningham, B. B., *Microchem. J. Symp. Ser.* **1**, 55 (1961).
26. Cunningham, B. B., *Colloq. Int. CNRS* **154**, 45 (1966).
27. Cunningham, B. B., *Proc. Int. Protactinium Conf., 3rd, Schloss Elmau, 1969*, BMBW-FBK 71–17, 14.1, 1971.
28. Cunningham, B. B., and Parsons, T. C., Doc. No. UCRL-20426, p. 239, University of California, Lawrence Radiation Laboratory, U.S. Atomic Energy Commission, Berkeley, CA, 1971.
29. Cunningham, B. B., and Wallmann, J. C., *J. Inorg. Nucl. Chem.* **26**, 271 (1964).
30. Damien, D. A., Haire, R. G., and Peterson, J. R., *J. Phys. Colloq.* **40**, C4–95 (1979).
31. Damien, D., Haire, R. G., and Peterson, J. R., *J. Less-Common Met.* **68**, 159 (1979).
32. Damien, D., Haire, R. G., and Peterson, J. R., *Inorg. Nucl. Chem. Lett.* **16**, 537 (1980).
33. Damien, M. D., Rep. CEA-N-816, p. 374, French Atomic Energy Commission, Fontenay-aux-Roses, 1967.
34. Dod, R. L., Ph.D. Thesis, University of California; Doc. No. LBL-659, Lawrence Berkeley Laboratory, U.S. Atomic Energy Commission, Berkeley, CA, 1972.
35. Eldred, V. W., and Curtis, G. C., *Nature (London)* **179**, 910 (1957).
36. Erdmann, B., and Keller, C., *Inorg. Nucl. Chem. Lett.* **7**, 675 (1971); *J. Solid State Chem.* **7**, 40 (1973).
37. Eubanks, I. D., and Thompson, M. C., *Inorg. Nucl. Chem. Lett.* **5**, 187 (1969).
38. Farr, J. D., Giorgi, A. L., Bowman, M. G., and Money, R. K., *J. Inorg. Nucl. Chem.* **18**, 42 (1961).
39. Fisher, E. S., *Trans. Met. Soc. AIME* **209**, 882 (1957).
40. Fowler, R. D., Matthias, B. T., Asprey, L. B., Hill, H. H., Lindsay, J. D. G., Olsen, C. E., and White, R. W., *Phys. Rev. Lett.* **15**, 860 (1965).
41. Fried, S., and Davidson, N., *J. Am. Chem. Soc.* **70**, 3539 (1948).
42. Fried, S., Westrum, E. F., Baumbach, H. L., and Kirk, P. L., *J. Inorg. Nucl. Chem.* **5**, 182 (1958).
43. Fuger, J., Peterson, J. R., Stevenson, J. N., Noé, M., and Haire, R. G., *J. Inorg. Nucl. Chem.* **37**, 1725 (1975).
44. Fuger, J., Haire, R. G., and Peterson, J. R., *J. Inorg. Nucl. Chem.* **43**, 3209 (1981).
45. Fuger, J., Haire, R. G., and Peterson, J. R., *J. Less-Common Metals* **98**, 315 (1984).
46. Fujita, D. K., Ph.D. Thesis, University of California; Doc. No. UCRL-19507, Lawrence Radiation Laboratory, U.S. Atomic Energy Commission, Berkeley, CA, 1969.
47. Goodman, L. S., Diamond, H., Stanton, H. E., and Fred, M. S., *Phys. Rev. A* **4**, 473 (1971).
48. Graf, P., Cunningham, B. B., Dauben, C. H., Wallmann, J. C., Templeton, D. H., and Ruben, H., *J. Am. Chem. Soc.* **78**, 2340 (1956).
49. Greiner, J. D., Peterson, D. T., and Smith, J. F., *J. Appl. Phys.* **48**, 3357 (1977).
50. Grosse, A. V., *J. Am. Chem. Soc.* **56**, 2200 (1934).
51. Gschneidner, K. A., *in* "Science and Technology of Rare Earth Materials" (E. C. Subbarao and W. E. Wallace, eds.), p. 25. Academic Press, New York, 1980.
52. Gschneidner, K. A., *in* "The Rare Earths in Modern Science and Technology" (G. J. McCarthy, J. J. Rhyne, and H. B. Silber, eds.), Vol. 2, p. 13. Plenum, New York, 1980.

53. Haire, R. G., in "Actinides in Perspective" (N. M. Edelstein, ed.), p. 309. Pergamon, New York, 1982.
54. Haire, R. G., Doc. No. ORNL-5817, p. 93, Union Carbide Corp., Oak Ridge National Laboratory, U.S. Department of Energy, Oak Ridge, TN, 1982.
55. Haire, R. G., and Asprey, L. B., *Inorg. Nucl. Chem. Lett.* **12,** 73 (1976).
56. Haire, R. G., and Baybarz, R. D., *J. Inorg. Nucl. Chem.* **36,** 1295 (1974).
57. Haire, R. G., and Baybarz, R. D., *J. Phys. Colloq.* **40,** C4-101 (1979).
58. Haire, R. G., Benedict, U., Peterson, J. R., Dufour, C., and Itié, J. P., *J. Less-Common Met.* **109,** 71 (1985).
59. Haire, R. G., Peterson, J. R., Benedict, U., and Dufour, C., *J. Less-Common Met.* **102,** 119 (1984).
60. Harbur, D. R., Anderson, J. W., and Maraman, W. J., *Mod. Cast.* **53**(3), 80 (1968).
61. Hübener, S., and Zvára, I., *Radiochim. Acta* **31,** 89 (1982).
62. Johnson, K. W. R., Doc. No. LA-1680, University of California, Los Alamos Scientific Laboratory, U.S. Atomic Energy Commission, Los Alamos, NM, 1954.
63. Johnson, K. W. R., and Leary, J. A., Doc. No. LA-2992, University of California, Los Alamos Scientific Laboratory, U.S. Atomic Energy Commission, Los Alamos, NM, 1964.
64. Jones, D. W., Farrant, S. P., Fort, D., and Jordan, R. G., in "Rare Earths and Actinides 1977" (W. D. Corner and B. K. Tanner, eds.), p. 11. Institute of Physics, London, 1978.
65. Kewish, R. W., Bard, R. J., Bertino, J. P., Fry, O. E., Hayter, S. W., Hill, F. J., Kelchner, B. L., and Savage, A. W., Jr., *Trans. Met. Soc. AIME* **215,** 425 (1959).
66. Kleinschmidt, P. D., Ward, J. W., Matlack, G. M., and Haire, R. G., *J. Chem. Phys.* **81,** 473 (1984).
67. Klemm, W., and Bommer, H., *Z. Anorg. Allg. Chem.* **231,** 138 (1937).
68. Lee, J. A., *Prog. Nucl. Energy, Ser. 5* **3,** 453 (1961).
69. Lesser, R., in "Gmelin Handbuch der Anorganischen Chemie" (G. Koch, ed.), System No. 71, Vol. 31, Part B1, p. 9. Springer-Verlag, Berlin and New York, 1976.
70. Liptai, R. G., and Friddle, R. J., *J. Cryst. Growth* **5,** 216 (1969).
71. Liptai, R. G., Lloyd, L. T., and Friddle, R. J., *Cryst. Growth Proc. Int. Conf. 1966*, p. 573 (1967).
72. Lohr, H. R., and Cunningham, B. B., *J. Am. Chem. Soc.* **73,** 2025 (1951).
73. Marples, J. A. C., *Acta Crystallogr.* **18,** 815 (1965).
74. Marples, J. A. C., *Colloq. Int. CNRS* **154,** 39 (1966).
75. Martinot, L., paper presented at the Journee d'etude des sels fondus, Soc. Chim. de Belgique, Liège, May 23-25, 1984.
76. McKay, H. A. C., Nairn, J. S., and Waldron, M. B., *Proc. UN Int. Conf. Peaceful Uses At. Energy 2nd* **28,** 299 (1958).
77. McWhan, D. B., Cunningham, B. B., and Wallmann, J. C., *J. Inorg. Nucl. Chem.* **24,** 1025 (1962).
78. McWhan, D., Montgomery, P. W., Stromberg, H. D., and Jura, G., Doc. No. UCRL-9808, University of California, Lawrence Radiation Laboratory, U.S. Atomic Energy Commission, Berkeley, CA, 1961.
79. McWhan, D. B., Wallmann, J. C., Cunningham, B. B., Asprey, L. B., Ellinger, F. H., and Zachariasen, W. H., *J. Inorg. Nucl. Chem.* **15,** 185 (1960).
80. Morgan, A. N., Johnson, K. W. R., and Leary, J. A., Doc. No. LAMS-2756, University of California, Los Alamos Scientific Laboratory, U.S. Atomic Energy Commission, Los Alamos, NM, 1962.

81. Mowat, J. A. S., and Yuille, W. D., *J. Less-Common Met.* **6**, 295 (1964).
82. Müller, W., Reul, J., and Spirlet, J. C., *ATW Atomwirtsch. Atomtech.* **17**, 415 (1972).
83. Müller, W., Reul, J., and Spirlet, J. C., *Rev. Chim. Miner.* **14**, 212 (1977).
84. Mullins, L. J., and Foxx, C. L., Doc. No. LA-9073, University of California, Los Alamos National Laboratory, U.S. Department of Energy, Los Alamos, NM, 1982.
85. Noé, M., and Peterson, J. R., *Transplutonium 1975, Proc. Int. Transplutonium Elem. Symp. 4th, 1975*, p. 69 (1976).
86. Okress, E. C., Wroughton, D. M., Comenetz, G., Brace, P. H., and Kelly, J. C. R., *J. Appl. Phys.* **23**, 545 (1952).
87. Peterson, D. T., Krupp, W. E., and Schmidt, F. A., *J. Less-Common Met.* **7**, 288 (1964).
88. Peterson, D. T., Schmidt, F. A., and Verhoeven, J. D., *Trans. Met. Soc. AIME* **236**, 1311 (1966).
89. Peterson, D. T., Page, D. F., Rump, R. B., and Finnemore, D. K., *Phys. Rev.* **153**, 701 (1967).
90. Peterson, D. T., and Schmidt, F. A., *J. Less-Common Met.* **24**, 223 (1971).
91. Peterson, J. R., Benedict, U., Dufour, C., Birkel, I., and Haire, R. G., *J. Less-Common Met.* **93**, 353 (1983).
92. Peterson, J. R., Fahey, J. A., and Baybarz, R. D., *Nucl. Metall.* **17**, 20 (1970).
93. Peterson, J. R., Fahey, J. A., and Baybarz, R. D., *J. Inorg. Nucl. Chem.* **33**, 3345 (1971).
94. See photomicrograph of this first isolated bulk sample of Bk metal in *Adv. Inorg. Chem. Radiochem.* **28**, 42 (1984).
95. Polonis, D. H., Butters, R. G., and Parr, J. G., *Research (London)* **7**, 273 (1954).
96. *Plutonium 1960, Proc. Int. Conf. Plutonium Metall. 2nd, 1960* (1961).
97. Raschella, D. L., Haire, R. G., and Peterson, J. R., *Radiochim. Acta* **30**, 41 (1982).
98. Schmidt, F. A., Lunde, B. K., and Williams, D. E., Doc. No. IS-4125, Iowa State University, Ames National Laboratory, U.S. Energy Research and Development Administration, Ames, IA, 1976.
99. Schmidt, F. A., Outlaw, R. A., and Lunde, B. K., Doc. No. IS-M-171, Iowa State University, Ames National Laboratory, U.S. Department of Energy, Ames, IA, 1979.
100. Schmidt, F. A., Outlaw, R. A., and Lunde, B. K., *J. Electrochem. Soc.* **126**, 1811 (1979).
101. Schmidt, F. A., Lunde, B. K., and Outlaw, R. A., *J. Spacecr. Rockets* **17**, 383 (1980).
102. Sellers, P. A., Fried, S., Elson, R. E., and Zachariasen, W. H., *J. Am. Chem. Soc.* **76**, 5935 (1954).
103. Shuck, A. B., *Plutonium React. Fuel, Proc. Symp.* 1967, p. 221 (1967).
104. Smith, J. F., Carlson, O. N., Peterson, D. T., and Scott, T. E., "Thorium: Preparation and Properties." Iowa State Univ. Press, Ames, 1975.
105. Smith, T. F., and Fisher, E. S., *J. Low Temp. Phys.* **12**, 631 (1973).
106. Spirlet, J. C., Rep. No. EUR 5412f, Commission of the European Communities, 1975.
107. Spirlet, J. C., *J. Phys. Colloq.* **40**, C4-87 (1979).
108. Spirlet, J. C., in "Actinides in Perspective" (N. M. Edelstein, ed.), p. 361. Pergamon, New York, 1982.
109. Spirlet, J. C., Bednarczyk, E., and Müller, W., *J. Less-Common Met.* **92**, L27 (1983).
110. Spirlet, J. C., and Müller, W., *J. Less-Common Met.* **31**, 35 (1973).
111. Spirlet, J. C., Müller, W., and van Audenhove, *J. Nucl. Instrum. Methods Phys. Res.* **A236**, 500 (1985).
112. Spirlet, J. C., and Vogt, O., in "Handbook on the Physics and Chemistry of the Actinides" (A. J. Freeman and G. H. Lander, eds.), p. 79. North-Holland Publ., Amsterdam, 1984.
113. Stevenson, J. N., and Peterson, J. R., *Microchem. J.* **20**, 213 (1975).
114. Stevenson, J. N., and Peterson, J. R., *J. Less-Common Met.* **66**, 201 (1979).

115. Stites, J. G., Salutsky, M. L., and Stone, B. D., *J. Am. Chem. Soc.* **77,** 237 (1955).
116. Thorsen, A. C., Joseph, A. S., and Valby, L. E., *Phys. Rev.* **162,** 574 (1967).
117. Wade, W. Z., and Wolf, T., *J. Inorg. Nucl. Chem.* **29,** 2577 (1967).
118. Wade, W. Z., and Wolf, T., *J. Nucl. Sci. Technol.* **6,** 402 (1969).
119. Wallmann, J. C., Crane, W. W. T., and Cunningham, B. B., *J. Am. Chem. Soc.* **73,** 493 (1951).
120. Ward, J. W., Kleinschmidt, P. D., and Haire, R. G., *J. Phys. Colloq.* **40,** C4–233 (1979).
121. Warner, J. C., *in* "Metallurgy of Uranium and its Alloys" (J. C. Warner, J. Chipman, and F. H. Spedding, eds.), Vol. 12A, p. 25. National Nuclear Energy Series, Division IV, 1953.
122. Westrum, E. F., Doc. No. MB-IP-96, University of California, Berkeley Radiation Laboratory, Berkeley, California, 1946. (Unavailable report referred to in reference *123* below).
123. Westrum, E. F., and Eyring, L., *J. Am. Chem. Soc.* **73,** 3396 (1951).
124. Westrum, E. F., and Eyring, L., *J. Am. Chem. Soc.* **73,** 3399 (1951).

ASTATINE: ITS ORGANONUCLEAR CHEMISTRY AND BIOMEDICAL APPLICATIONS

I. BROWN

The Research Laboratories, The Radiotherapeutic Centre,
University of Cambridge School of Clinical Medicine,
Addenbrooke's Hospital, Cambridge CB2 2QQ, England

I. Introduction

Astatine ($^{200-219}_{85}$At), the fifth and heaviest member of the Periodic Table Group VIIB (the halogens), is the earth's rarest naturally occurring element. All its isotopes are radioactive (Table I), hence the Greek name αστατωζ, meaning *unstable* (44, 45). The possibility of their existence was predicted from the β-decay of polonium (55). Its three naturally occurring isotopes, ^{215}At, ^{218}At, and ^{219}At, are the extremely short-lived natural daughters of AcA (77), RaA (76, 173), and AcK (72), respectively.

The mass–energy dependence for heavy nuclei shows that, in element $Z = 85$, the proton:neutron ratio is such that nuclear stability is not expected. The most long-lived isotopes possess a number of neutrons close to the magic number of 126, corresponding to the formation of a closed nuclear shell configuration (20, 75). The relationship between the emitted α-particle energy and mass number of the astatine isotope is shown in Fig. 1. The lightest isotopes have very short half-lives and disintegrate with the expulsion of high-energy α-particles, as would be expected since the products of α-emission have a next-to-magic-number of protons. Moreover, in possessing a considerable neutron deficiency, these isotopes would certainly be unstable upon capture of an orbital electron. On passing the neutron-surplus region $(A - Z = 126)$ the curve shows a break, which reveals a sharp decrease in the strength of bonding of nucleons within the nucleus. All the isotopes with $A > 211$ have very short half-lives; the heaviest isotopes $\geq ^{219}$At show some stability toward α-decay. However, in this range transitions by β-decay are also possible (177, 118). It has been suggested that astatine, like technetium and promethium, cannot have β-stable nuclei at all (51).

TABLE I

Decay, Half-Lives, and Production of Astatine Isotopes[a]

Isotope	Half-life[b]	Decay	Production
^{219}At	0.9 min	97% α, 3% β	Natural daughter of AcK
^{218}At	1.5–2.0 s	99.9% α, 0.1% β	Natural daughter of RaA
^{217}At	0.032 s	α	Daughter of ^{221}Fr in ^{233}U series
^{216}At	3×10^{-4} s	α	Daughter of ^{220}Fr in ^{228}Pa series
^{215}At	1×10^{-4} s	α	Daughter of ^{219}Fr in ^{227}Pa series
^{214}At	Short (2×10^{-6} s)	α	Daughter of ^{218}Fr in ^{228}Pa series
^{213}At	0.11×10^{-6} s	α	Decay product of ^{225}Pa
^{212}At	0.315 s	α	Bi(α,n)
^{212m}At	0.12 s	α	Bi(α,n)
^{211}At	7.21 h	40.9% α, 59.1% EC	Bi(α,2n); ^{211}Rn (EC)
^{210}At	8.3 h	99.9% EC, 0.1% α	Bi(α,3n); ^{210}Rn (EC)
^{209}At	5.4 h	95% EC, 5% α	Bi(α,4n); ^{209}Rn (EC)
^{208}At	1.63 h	99% EC, 0.5% α	Bi(α,5n); daughter of ^{212}Fr
^{207}At	1.8 h	α, EC	Bi(α,6n); Au + ^{14}N; ^{207}Rn (EC)
^{206}At	31 min	EC, α	Bi(α,7n); Au + ^{12}C, ^{14}N, ^{16}O
^{205}At	26 min	α, EC	Bi(α,8n); Au + ^{12}C, ^{14}N, ^{16}O Pt + ^{14}N
^{204}At	9.1 min	EC, α	Bi(α,9n); Au + ^{12}C, ^{14}N, ^{16}O
^{203}At	7.3 min	α, EC	Bi(α,10n); Pt + ^{14}N; Au + ^{12}C, ^{14}N, ^{16}O
^{202}At	3.0 min	α, EC	Au + ^{12}C, ^{14}N, ^{16}O
^{201}At	1.5 min	α	Au + ^{12}C, ^{14}N, ^{16}O
^{200}At	0.8 min	α	Au + ^{12}C

[a] All astatine isotopes, with the exception of ^{213}At, produce other radionuclides by their decay, consequently complicated decay curves can arise. In astatine isotopes, electron capture (EC) always produces K-radiation.

[b] h, Hours; min, minutes; s, seconds.

Many aspects of the nuclear physics, and inorganic and organic chemistry, of the astatine radionuclides have been the subject of a number of excellent reviews (5, 6, 17, 18, 49, 79, 80, 90, 110).

II. Isotopes: Production, Extraction, and Identification

Preparation of the isotopes of astatine is more difficult than with most radionuclides, as they cannot be synthesized by neutron irradiation; this precludes the use of a nuclear reactor. To date, the bulk of

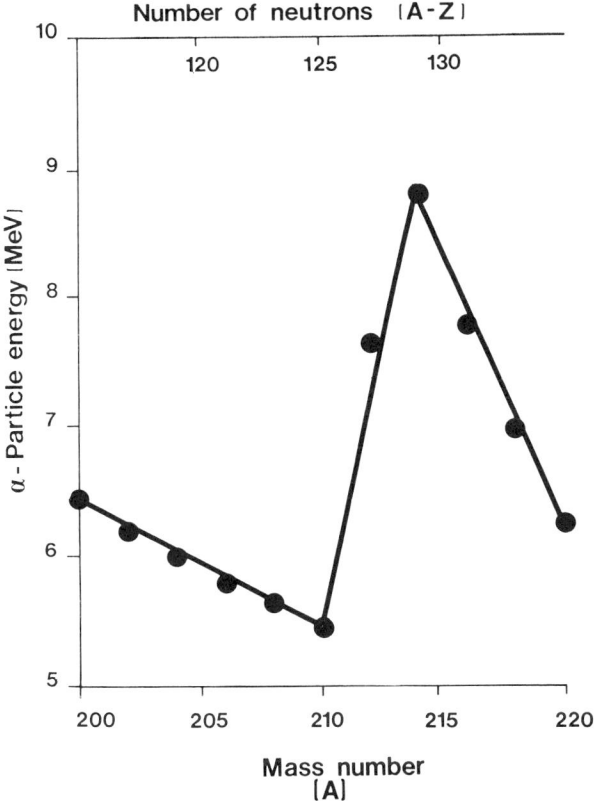

FIG. 1. Relationship between the energy of the α-particles emitted by astatine isotopes and their mass number (A) and neutron content (A − Z).

astatine research has been carried out using cyclotrons or synchrocyclotrons at large nuclear research centers.

Numerous nuclear reactions have been employed to produce astatine. Three of these are particularly suited for routine preparation of the relatively long-lived isotopes with mass numbers 209, 210, and 211. The most frequently used is the $^{209}\text{Bi}(\alpha,xn)^{213-x}\text{At}$ ($x = 1$–4) reaction, in which bismuth (44, 74, 120) or bismuth oxide (7, 125) is bombarded by 21- to 40-MeV α-particles. The $^{209}\text{Bi}(\text{He}^+,xn)^{213-x}\text{At}$ reaction can also be used to produce isotopes of astatine (152); the nuclear excitation functions (62) favor a predominant yield of ^{209}At and ^{210}At. The routine preparation of astatine is most conveniently carried out through the $^{209}\text{Bi}(\alpha,xn)^{213-x}\text{At}$ nuclear reactions, from which a limited spectrum of astatine nuclides may be derived. The excitation functions for these nuclear reactions have been studied extensively (78, 89, 120). The incident α-particle threshold energies for (α,2n), (α,3n) and (α,4n) reactions correspond to 21, 29, and 34 MeV, respectively (see Fig. 2). For routine chemical studies with astatine, an isotopic mixture will suffice. Large quantities of astatine isotopes can be synthesized by using both internal (*100, 129*) and external target systems (*22*). Activities up to 1 Ci ml^{-1} have been produced using an internal target (*48*). However, for biomedical studies, only. ^{211}At possesses advantageous decay properties; ^{210}At decays by electron capture to ^{210}Po, which subsequently decays by the emission of 5.355- to 5.519-MeV α-particles, with a half-

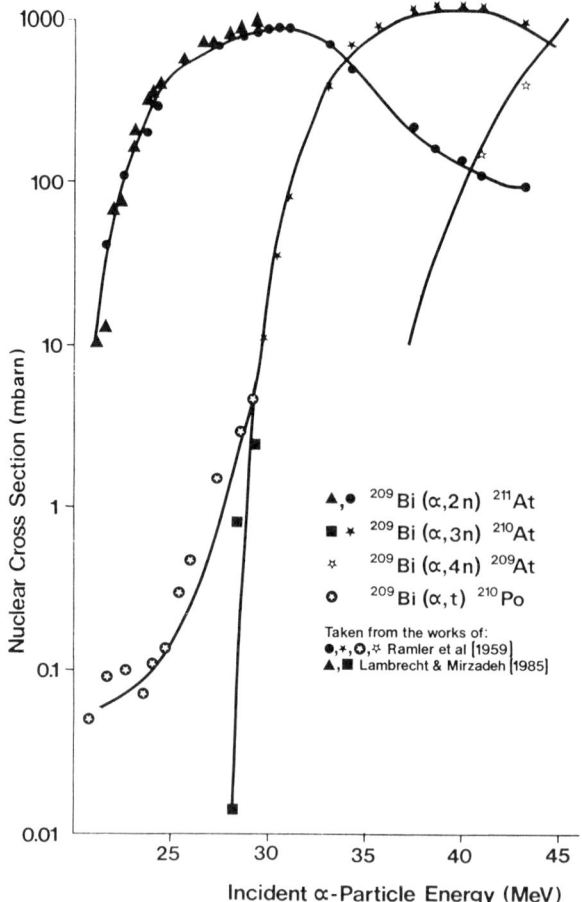

FIG. 2. Excitation functions for the production of ^{209}At–^{211}At by α-particle bombardment of ^{209}Bi.

life of 138.4 days. Thus ^{210}At clearly represents a potentially significant radiation hazard, particularly if ingested. Astatine-211 can be exclusively produced by the irradiation of bismuth with 28-MeV α-particles, where the (α, 2n) nuclear reaction cross-section (σ) is near its optimum at 790 mbarn. It has been possible to prepare the high-activity yields of ^{211}At potentially required for biomedical application by using an external oscillating α-beam cyclotron target system which also incorporates an almost 4π combined helium/water target cooling (*33*); beam currents up to 30 μA have been routinely employed.

Astatine can be readily and routinely obtained in an inorganic form suitable for chemical and biomedical application from the α-irradiated bismuth target by either extractive (*7, 113, 116*) or dry (*2, 3, 48, 74, 93, 120*) distillation techniques. Both methods have their relative merits (*33, 101*).

The wet technique involves dissolution of the bismuth target in a mineral acid such as HNO_3 or $HClO_4$, followed by extraction of astatine in its highly reactive zero-oxidation state with an organic solvent such as isopropyl ether. The back-extraction of astatine from the organic layer can be achieved with NaOH containing a little Fe^{3+}; $Fe(OH)_3$ retains the impurities and the loss of At does not exceed 5%. The yield has been reported at ~90% (*116*). Alternatively, distillation of the separated organic layer enables At^0 to be carried over, either via adsorption or formation of adducts, into a cold trap. The final valence of astatine can easily be determined: Oxidation to At^+ or reduction to At^- can be achieved by addition of 1 M $HNO_3/10^{-3}$ M $H_2Cr_2O_7$ or by aqueous 0.2 M Na_2SO_3, respectively. The organic solvent is removed by distillation through the nonpolar/polar media. Yields of approximately 60% have been obtained for At^-, At^+, and At^0 (*101*) and high activities can be obtained, though the method is more suitable for low-activity requirements.

Distillation of astatine from molten metallic bismuth is the most widely used technique for separation of the radionuclide. In order to separate astatine from the mass of bismuth, lead, and polonium, use is frequently made of its high volatility. Distillations have been carried out at temperatures ranging from 300 to 600°C, in a stream of N_2 or He (*11*), or under *in vacuo* (*74, 78*) conditions. Astatine has been collected by direct adsorption from the gas phase onto platinum or silver (*2, 3, 48, 78*), condensed on a glass cold finger probe or into frozen solutions at the temperature of liquid N_2 (*101*), as well as absorbed into aqueous solutions of NaOH or Na_2SO_3 (*1*). More recently, high yields of $^{211}At^-$ have been obtained by trapping distilled At^0 on a silica–gel column, then eluting with aqueous NaOH into a small volume (*33, 89*).

Meyer and Rössler (*101*) showed that the overall yields for wet and dry extractive procedures are comparable, being approximately 60% for $^{211}At^-$. Some workers have found that yields vary somewhat, due to adsorption of evaporated At^0 onto vessel walls, and to the possibility of the retention of astatine within the target due to the formation of nonvolatile compounds (*11*). However, the dry evaporation method is more applicable to studies with high-activity targets; it is rapid and lends itself to further development within the scope of remote handling techniques. Aspects of both extraction approaches have been discussed widely (*2, 7, 33, 89, 101, 116, 120, 160*).

Another method employed to prepare astatine uses secondary nuclear reactions and spallation processes. Heavy metals such as lead, bismuth, thorium, and uranium can be bombarded with high-energy particles, such as 160-MeV protons, 190- to 270-MeV Ar^+, and 430- to 500-MeV Kr^+ (*23, 24, 91, 92*). These processes lead to a wide range of radionuclides, including some astatine isotopes. Fission of thorium can also be achieved by 660-MeV protons (*16, 28, 85*). However, the formation cross-section for the spectrum of astatine isotopes is a factor of two to three times less than the total formation cross-section for products of spallation and fission (*177*). Radiochemically pure astatine can be obtained only by its separation from most elements of the periodic table. After dissolving the target material, astatine can be separated and purified by repeated adsorption on, and elution from, tellurium-filled columns (*79, 174*). More recently, its isolation from other spallation products has been achieved by the simple and elegant technique of gas thermochromatography (*98*).

High-energy nuclear reactions can similarly be utilized to synthesize the neutron-deficient noble gas isotopes among the spectrum of spallation products. Isolation of the radon isotopes from the mixture is easily performed by gas chromatography in columns packed with molecular sieves (*82*). The longest lived of the radon isotopes, ^{211}Rn ($\tau_{1/2} = 14.6$ hours), decays by electron capture to ^{211}At. Radon and astatine isotopes can also be obtained by heavy-ion induced nuclear reactions (*52, 99*) or photospallation (*172*). Particularly promising is the production of ^{211}Rn, and hence ^{211}At, by the $^{209}Bi(^7Li,5n)^{211}Rn$ reaction, using 53- to 60-MeV ^7Li ions ($\sigma = 500$-600 mbarns) accelerated from a three-stage Tandem Van der Graaff system (*103*). In contrast to spallation reactions on uranium and thorium, which have been used for producing ^{211}Rn, the other competing nuclear reactions ($^7Li,xn$), ($^7Li,pxn$), and ($^7Li,\alpha xn$) (where $x = 1$-4) lead to either very short-lived radon, astatine, polonium, bismuth, and lead isotopes, which decay by α-emission, or to radionuclides with a considerably longer half-life. The limited product spectrum allows the effective separation of ^{211}Rn by a simple degassing of the bismuth target at 500°C and trapping of the products in silver-wool and tellurium filters. Radon-211 can be obtained with high radiochemical purity, is a convenient generator for ^{211}At (Scheme 1), and delivers its maximum activity after 14 hours. About 50% of the maximum ^{211}At activity is, however, still available after 40 hours (*99*).

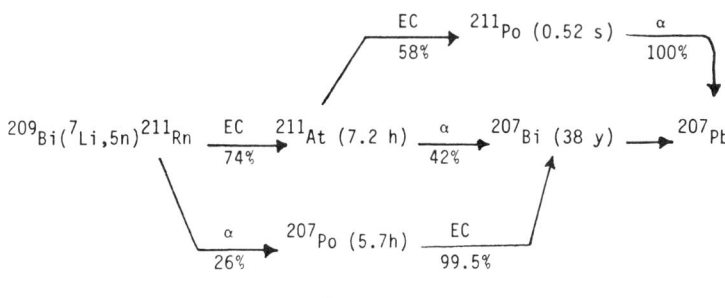

SCHEME 1

Identification and quantitation of the three main astatine isotopes, ^{209}At, ^{210}At, and ^{211}At, can be achieved by the appropriate measurement of α-, γ-, or X-ray activity. X-rays and γ-rays can be counted conveniently by well-type crystal counters, whereas α-counting[1] requires that astatine samples be measured as infinitely thin or thick preparations. Astatine-211 can be measured by counting its 79- to 92-keV ^{211}Po $K-L,M,N$ X-rays, and, importantly, discriminated from ^{210}At and ^{209}At by their γ-emissions (245, 1180; 195, 545, and 782 keV, respectively). Identification and aspects of counting other astatine isotopes have been briefly discussed elsewhere (*6, 110*).

III. Synthetic Organic Radiochemistry

Astatine, generally speaking, is a difficult isotope to study from a chemical viewpoint because no stable isotopes exist. Although the study of the chemical properties of astatine began over 40 years ago (*44*), the element's precise behavior is still in doubt. The chemical similarity between astatine and its nearest halogenic neighbor, iodine, is not always obvious. In many cases the astatine tracer has not

---

[1] α-Emissions arising from long-lived ^{210}Po may complicate determinations, but if ^{211}At is allow to decay completely (>99.9% in 48 hours), discrimination can be achieved.

followed the iodine carrier as would be expected. The situation is further complicated by the fact that the chemical properties of the halogens do not follow an easily recognizable pattern; in many respects bromine does not interpolate between chlorine and iodine (49). However, the general trend in the periodic system suggests that astatine is more metallic in character than iodine.

The chemistry of ^{211}At, as a consequence of its production by nuclear reaction processes and its short physical half-life, must invariably be carried out on a tracer scale. At high concentration, the intensive accompanying α-particle radiation would give rise, in aqueous solutions, to labile and highly reactive peroxides, and to heat that would interfere with the understanding and elucidation of reaction stoichiometry, due to the production of very short-lived metastable radiolytic species. The concentration of usable ^{211}At is clearly limited by these factors. For ^{211}At, whose specific activity is 1.5×10^{16} α-particles min^{-1} cm^{-3}, a $1M$ ^{211}At solution would lead to unacceptably intense radiation dose and heat. The highest concentration of ^{211}At that has been prepared is 10^{-8} M (2, 3); that needed for typical chemical experiments is in the range 10^{-13}–10^{-15} M. The concentration of ^{211}At in 1 ml of a solution with an activity of 40 μCi is only 10^{-10} M or 0.002 ppb. Therefore, extremely pure reagents are required, otherwise reactions with impurities may lead to severe irreproducibility, and incomprehensible chemical behavior and identification problems.

The low concentration of ^{211}At has several consequences important in the understanding of its chemistry.

1. The diatomic molecule At$_2$, although purported to have been identified (115), is on statistical grounds unlikely to exist, and in terms of organic syntheses there would be a negligible probability of diastatination occurring.

2. Equilibrium reactions such as

$$R^- + At^+ \rightleftharpoons RAt \rightleftharpoons R^+ + At^- \qquad (1)$$

do not occur, because if the At is removed from its substrate R, the reverse reaction with the same substrate is statistically highly improbable. If it occurs, complete decomposition of an At compound can be considered irreversible.

3. Disproportionation, a common process in inorganic halogen chemistry, does not proceed. For example, IOH(HIO) is an unstable, strongly electrophilic compound and, when no substrate is present,

readily disproportionates:

$$3IOH \longrightarrow 2HI + HIO_3 \qquad (2)$$

At low concentrations such a disproportionation would be virtually impossible. Therefore, when AtOH is formed it will have a relatively long lifetime.

4. Normal physicoorganic methods used for the formal identification of organic compounds are not applicable to organic astatine chemistry. The mass quantities required for the characterization of compounds by UV, NMR, and IR spectroscopy are in the region 10^{-6}–10^{-3} g; molar concentrations of $\sim 10^{-11}$ preclude the application of such techniques. Mass spectrometry has not yet been developed to operate at such a concentration, except under special laboratory conditions (4).

The only method that can be used routinely to identify organoastatine compounds is measurement of radioactivity based upon its distribution over two or more phases. Such techniques are gas–liquid chromatography (GLC), high-pressure liquid chromatography (HPLC), thin-layer chromatography (TLC), and electrophoresis.

In order to identify synthesized astatocompounds it has been necessary to adopt criteria and guidelines for both synthesis and identification.

1. A series of analogous, well-characterized iodo compounds should be prepared under conditions identical to those used for astatocompounds.

2. Syntheses should be carried out under the mildest conditions possible; the use of drastic and strongly oxidative conditions may lead to ill-defined reaction mixtures.

3. Wherever possible, specific astatocompounds should be synthesized by different radiochemical routes (128).

4. The chromatographic or electrophoretic behavior of a specific astatocompound must be considered similar, but not necessarily the same, as that of its analogous iodo and bromo derivative. This procedure has been developed and termed *sequential analysis* (99, 102); in order to avoid a coincidental fitting of chromatographic data, several solvent systems should be used.

5. The physical and chemical behavior of astatocompounds must be considered comparable to the corresponding lower halogenic compounds.

The adoption of such criteria has ensured that more consistent and reproducible results are obtained in synthetic inorganic and organic astatine chemical studies (*17, 18, 33, 99, 160*).

A. COMPOUNDS OF MULTIVALENT ASTATINE

As astatine exhibits a more pronounced positive character compared with the other halogens, early studies were directed toward the preparation of organometallic derivatives of astatine in the valence states +3 and +5. The compounds $RAtCl_2$, R_2AtCl, and $RAtO_2$ (where $R = C_6H_5$ or $p\text{-}CH_3C_6H_4$) were prepared according to the reaction schemes in Eq. (3); macroquantities of iodine carrier labeled with ^{131}I were always added (*112, 114*). Consequently, formation of the analogous

$$R_2ICl + At^- \longrightarrow Cl^- + R_2IAt \xrightarrow{170°C} RAt + RI \xrightarrow[Cl_2]{0°C} RAtCl_2 \qquad (3)$$

iodine compounds along with those of the astato compounds occurred at each stage during the synthesis. In order to prepare $RAtCl_2$, KI containing At^- was added to a solution of R_2ICl. The crystalline product R_2I_2 (R_2IAt) was centrifuged, washed, and heated in sealed glass ampuls for a few minutes at 170–190°C. The thermal decomposition product RI(RAt) was dissolved in $CHCl_3$ and chlorinated at 0°C in order to obtain $RICl_2(RAtCl_2)$. The analogous iodo compound was present as a yellow precipitate which was crystallized from $CHCl_3$ together with the product. To a hot $CHCl_3$ solution of $RAtCl_2$ was slowly added R_2Hg [see Eqs. (4–6].

$$RAtCl_2 + R_2Hg \longrightarrow R_2AtCl + RHgCl \qquad (4)$$

$$RAtCl_2 + RHgCl \longrightarrow R_2AtCl + HgCl_2 \qquad (5)$$

$$RAtCl_2 + OCl^- + 2OH^- \longrightarrow RAtO_2 + 3Cl^- + H_2O \qquad (6)$$

After cooling, $HgCl_2$ precipitated out leaving a mixture of $RAtCl_2$ and R_2AtCl, together with their iodo analogs, in solution. The extraction of R_2AtCl was performed into an aqueous phase and identified by TLC. If a solution of NaOH and CH_3COOH is added to the crystals of $RICl_2(RAtCl_2)$, and the mixture heated to its boiling point and chlorinated in order to oxidize $I^{3+}(At^{3+})$ into $I^{5+}(At^{5+})$, then a white crystalline precipitate of $RIO_2(RAtO_2)$ formed on cooling. This was recrystallized from a small amount of water.

The carrier compounds were useful for identifying the corresponding astatine derivatives by means of paper chromatography (*112*) and TLC (*114*); the β- and α-activities of ^{131}I and ^{211}At products, respectively, were measured.

B. COMPOUNDS OF MONOVALENT ASTATINE

1. *Aliphatic Compounds*

a. Alkyl Astatides. The simplest molecule, the methyl astatide ion CH_3At^+, has been identified by time of flight mass spectrometry (*4*); the observation of its mass line was a result of an astatine reacting with organic impurities within the ion source. Methyl astatide is one of the products formed from recoil ^{211}At atoms, originating from the ^{211}Rn(EC)^{211}At nuclear transformation, reacting with various long-chain and cyclic aliphatic hydrocarbons, and benzene (*88*). In addition to the formation of CH_3At, the higher homologs, C_2H_5At, isomers of C_3H_7At, C_4H_9At, $C_6H_{13}At$, c-C_5H_9At, and c-$C_6H_{11}At$ were also produced as a result of a spectrum of ^{211}At recoil reactions with the parent hydrocarbons. The separation of reaction products and their identification was effected by GLC; the relative amounts of the individual products varied for different starting hydrocarbons and depended very much upon experimental conditions. However, as a rule, the highest yield was of the astatinated derivative of the original hydrocarbon (*88*).

Normal alkyl astatides n-$C_nH_{2n+1}At$ ($n = 2$–6) have been prepared by homogeneous halogen-exchange reactions (*127*); molecular iodine containing AtI was added to the corresponding n-alkyl iodides at room temperature according to Eq. (7), where $R = C_2H_5$–C_6H_{13}; ethyl

$$RI + At^- \longrightarrow RAt + I^- \tag{7}$$

alcohol accelerated the exchange reaction. Normal alkyl astatides have also been prepared by gas chromatographic halogen exchange, where At$^-$ had been previously adsorbed on the solid phase of a short precolumn packed with kieselguhr and the respective alkyl iodide was allowed to flow through the GLC column in a carrier gas stream. Halogen exchange occurred best at 130–200°C; products separation was achieved by using a longer sequential analytical chromatographic column (*125, 127*). Preparation of n-alkyl astatides ($n = 2$–5) and i-alkyl astatides, i-$C_nH_{2n+1}At$ ($n = 3$–5), has been achieved by a simplified GLC halogen-exchange method using only one analytical column with At$^-$ adsorbed at its inlet (*54*).

The boiling temperatures (T_b) for alkyl astatides obtained by halogen-exchange methods have been determined by extrapolation from the T_b values of the corresponding lighter alkyl halides using the technique of sequential GLC analysis (53, 83, 125, 127). Typical values are given in Table II; an extensive tabulation of extrapolated data has been collected by Berei and Vasáros (17, 18).

A similar linear dependence of T_b on the logarithmic GLC retention time for alkyl iodides and alkyl astatides has been observed (54). The calculation of an experimental parameter Z' (83) for astatine has also

TABLE II

EXTRAPOLATED PHYSICOCHEMICAL PROPERTIES OF ALIPHATIC ASTATIDES[a,b,c]

Compound	Mean T_b (°C)	H_{vap} (kJ mol^{-1})	D_{C-At} (kJ mol^{-1})	IP (eV)
CH$_3$At	66 ± 3 (86)	28.9 (115)	139.3 (115)	8.85 (115)
	65.8 (115)		205 (67)	8.8 (67)
	73 ± 5 (8)		176 (141)	
	72 ± 2 (66)			
	77 ± 5 (67)			
C$_2$H$_5$At	98 ± 2 (125, 127)	31.6 (115)	167 (141)	8.8 (115)
	95.4 (115)			8.65 (67)
	103 ± 5 (8)			
n-C$_3$H$_7$At	123 ± 2 (125, 127)		163 (141)	
	124.3 (115)		161.5 (139)	
i-C$_3$H$_7$At	112 ± 2 (54)		159 (141)	
	110.4 (115)		151.9 (139)	
n-C$_4$H$_9$At	152 ± 3 (125, 127)		161 (141)	
	152.5 (115)			
i-C$_4$H$_9$At	142 ± 3 (54)			
	140.4 (115)			
n-C$_5$H$_{11}$At	176 ± 3 (125, 127)			
	182 (115)			
i-C$_5$H$_{11}$At	163 ± 3 (125, 127)			
	163 (115)			
n-C$_6$H$_{13}$At	202 ± 2 (125, 127)			
CH$_2$AtCl	137 ⎫		130.1 ⎫	
CH$_2$AtBr	168 ⎬ (115)		124.7 ⎬ (115)	
CH$_2$AtI	208 ⎭		118.0 ⎭	

[a] References in parentheses.
[b] GLC data (54, 125, 127).
[c] Thermal decomposition data (139).

been calculated through studies dealing with the gas chromatographic behavior of alkyl astatides in relation to that of other alkyl halides. Using this derived parameter for extrapolation, T_b, heat of vaporization (ΔH_{vap}), ionization potential (IP), and C–At bond dissociation energies (D_{C-At}) have been estimated for a number of simple organoastatine molecules (115). The T_b, IP, and D_{C-At} for ethyl astatide have also been calculated by other extrapolation techniques, based upon the relationships between different physicochemical constants (8, 66, 67)[2]. Similar methods have also been used to calculate the D_{C-At} values for a number of alkyl astatides (141). More recently, the dissociation energy of the carbon–astatine bond in n- and i-propyl astatide has been established by measurement of the kinetics of the pyrolytic decomposition of these compounds (140); the values obtained were of a scale similar to those obtained by extrapolation (see Table II).

b. *Astatocarboxylic Acids.* The halogen atoms of haloacetic acids are readily replaced by heavier halogens in aqueous solution. Astatoacetic acid was prepared by reacting At⁻ in the presence of an iodide carrier with an aqueous solution of iodoacetic acid at 40°C (125, 126).

$$At^- + ICH_2COOH \longrightarrow AtCH_2COOH + I^- \qquad (8)$$

The product was extracted with ethyl ether and after evaporation of the solvent the dry residue was recrystallized from CCl_4; $AtCH_2COOH$ was identified by ion-exchange chromatography. Its dissociation constant was determined by the measurement of its distribution between diisopropyl ether and aqueous solutions of varying acidity; K_a was found to be 1.5–1.8×10^{-4} mol liter^{-1} between 0 and 27°C (125, 126).

2. Aromatic Compounds

a. *Astatobenzene.* Astatobenzene has been successfully synthesized by a number of well-established preparative routes (Table III). The

---

[2] In a study directed toward predicting physicochemical properties of volatile compounds of the superheavy elements (8), monotonic relationships between physicochemical constants were derived for alkyl derivatives of elements which belonged to the same group of the periodic system. A number of estimates were also made for the corresponding astatine derivatives. Since both the electronegativity (χ) and atomic volume (v_A) influence the van der Waal's interactions of their organic derivatives, T_b was plotted against $f(Zv_A\chi^{-1})$, where Z is the atomic number of element A. Smooth curves were obtained for methyl and ethyl halides (8); the values v_A and χ are themselves extrapolated from corresponding values for other halogens. Similarly derived expressions were related for IP [$f(Zr_A^2\chi^{-1})$] and D_{C-At} [$f(Z\chi^{-1})$], where r_A is the covalent atomic radius of A.

TABLE III Syntheses of Astatobenzene

Method	Substrate	Experimental conditions	Yield (%)	Reference
Electrophilic substitution (At$^+$)				
Homogeneous	C_6H_6	$CH_3COOH/HClO_4$ mixture containing $Cr_2O_7^{2-}$, heated to 120°C for 90 minutes	50	140
Heterogeneous	C_6H_6	$HClO_4$ or H_2SO_4 at 180–190°C for 20 minutes	90	148
"Electrophilic" halogen exchange (ipso-attack)				
$\overset{\delta+}{At}$—$\overset{\delta-}{Cl}$	C_6H_5Cl C_6H_5Br	AtCl, AtBr at 60°C for 1 hour	29, 19 45, 30	99, 104
Nucleophilic substitution (At$^-$)				
Homogeneous halogen exchange	C_6H_5I	At$^-$ (I_2)/γ-radiation	73	125, 128
	C_6H_5Cl			
	C_6H_5Br	n-$C_4H_9NH_2$ at 210°C	85	143, 144
	C_6H_5I	for 30–60 minutes	99	
	C_6H_5I	At$^-$ (KI), GLC synthesis/separation at 130–200°C		125, 128
Heterogeneous halogen exchange	C_6H_5Br	At$^-$ (NaI) at 155°C for 2 hours	60	84
	C_6H_5Br	At$^-$ (NaOH) sealed ampul at 250°C for 30 minutes	70	141, 142
	$(C_6H_5)_2I_2$	At$^-$ (KI)/hot C_2H_5OH, decomposition of crystallized product at 175°C in sealed tube		125, 128
Via diazonium intermediate	$C_6H_5NH_2$ $C_6H_5NH_2$ $C_6H_5NHNH_2$	Diazotization and decomposition in the presence of At$^-$/KI	80	69, 71 125, 128 125, 128
Recoil astatination [At]*	$(C_6H_5)_3Bi$	Via $^{209}Bi(\alpha,2n)^{211}At$		125, 128
	C_6H_6		44	88, 111
	C_6H_6	Via $^{211}Rn(EC)^{211}At$	23	150
	C_6H_5Cl		35	
	C_6H_5Br		40	
	C_6H_5I		45	
	C_6H_5Cl		60	145

production of astatobenzene in high yields has been achieved best through halogen-exchange (At for I or Br) reactions, in both homogeneous and heterogeneous systems. Electrophilic substitution can also be very successful particularly if the heterogeneous systems contain strong acids. Astatobenzene has been identified by GLC.

The electrophilic substitution of benzene by At^+ in a homogeneous mixture of $HClO_4$ and CH_3COOH containing $Cr_2O_7^{2-}$ as an oxidizing agent has resulted in good yields (*139*). The reactants, sealed within glass ampuls, were heated for 30–60 minutes at 100–120°C; astatobenzene was obtained in ~50% yield. More efficient electrophilic astatination has been achieved in heterogeneous systems with radiochemical yields of approximately 90% if astatine is present in the aqueous phase containing either H_2SO_4 or $HClO_4$, even in the absence of $Cr_2O_7^{2-}$ oxidizing agent (*148*). This latter finding has been interpreted on the basis of the complex structure[3] of the monovalent At^+ in aqueous solutions by the fact that, in the presence of strong acids, the equilibrium in Eq. (9) shifted to the right. It has been assumed that the

$$HOAt + H_3O^+ \rightleftharpoons [H_2OAt]^+ + H_2O \qquad (9)$$

heterolytic fission of hypoastatous acid $[H_2OAt]^+$, resulting in the formation of the reactive entity At^+, is the rate-determining step in the astatination mechanism, immediately preceeding its substitution into the aromatic ring. The activation energy (E_a) for the electrophilic astatination of benzene has been obtained from kinetic studies and found to be 134 ± 8 kJ mol^{-1} (*148*).

Electrophilic halogen-exchange reactions in monohalobenzenes by the interhalogen compounds AtCl and AtBr yielding astatobenzene have been described (*99, 104*). This unexpected observation has been explained by Meyer *et al.* (104) in terms of an "*ipso*-attack" of the highly polarized interhalogen molecule at the electronegative site of the substrate, and consequent complex formation (Scheme 2). This results in electrophilic replacement of the halogen atom competing with the normal aromatic substitution of hydrogen in the ortho position (*99, 104*). It has been assumed that both reactions are assisted by a Lewis

---

[3] A relatively stable aqua complex or protonated hypoastatous acid $[H_2OAt]^+$ has been assumed; similarly, a protonated hypoiodous acid has been reported to exist in aqueous solutions (*15*). The equilibrium constant for the deprotonation reaction [Eq. (9)] has been estimated by extrapolation of data accrued from the lighter halogens to be $<10^{-3}$ (*80*), indicating that $[H_2OAt]^+$ is a fairly weak acid. Another structure, the symmetric diaqua cationic complex $[H_2O-At-OH_2]^+$, has also been proposed (*79, 80*).

SCHEME 2. X = F, Cl, Br; B = Lewis Base.

base, which is always present in the reaction mixture. However, more recent studies have found that the yields for the halogen as well as for the hydrogen replacement are not reproducible; this is probably due to the ill-defined chemical state of astatine under the specified experimental conditions (43).

Astatobenzene has also been prepared by heterogeneous halogen-exchange reaction between At^- adsorbed on solid sodium iodide, and bromobenzene at its boiling point (84). A further development in this technique has been to allow the reaction of bromobenzene in sealed ampuls at 250°C with At^- adsorbed on sodium hydroxide; this resulted in high yields of about 70% (141, 142).

Nucleophilic astatination of halobenzenes C_6H_5X (X = Cl, Br, I) in homogeneous mixtures with $n\text{-}C_4H_9NH_2$, $(C_2H_5)_2NH$, and $(C_2H_5)_3N$ at 210°C has led to the formation of astatobenzene with radiochemical yields of 75–90% (143, 144). A two-step process has been postulated [Eqs. (10) and (11)], with the latter reaction [Eq. (11)] as the rate-determining step:

$$C_6H_5X + NRR'R'' \longrightarrow [C_6H_5NRR'R'']^+ X^- \qquad (10)$$

$$[C_6H_5NRR'R'']^+X^- + At^- \longrightarrow C_6H_5At + NRR'R'' + X^- \qquad (11)$$

where R, R', R'' = H, C_2H_5, or $n\text{-}C_4H_9$. Kinetic investigations have supported this assumption as the activation energy for halogen exchange remained the same regardless of the different halogen leaving groups. The reaction, however, is significantly influenced by the nature of the amine; this is probably related to steric effects (see Table IV).

TABLE IV

ACTIVATION ENERGIES FOR NUCLEOPHILIC
ASTATINATION OF HALOBENZENES:
THE INFLUENCE OF ALIPHATIC AMINES

Reaction	E_a (kJ mol^{-1})
At$^-$ + C$_6$H$_5$Br + (C$_2$H$_5$)$_3$N	111.7
At$^-$ + C$_6$H$_5$Br + (C$_2$H$_5$)$_2$NH	21.8
At$^-$ + C$_6$H$_5$Br + C$_4$H$_9$NH$_2$	17.2
At$^-$ + C$_6$H$_5$Cl + C$_4$H$_9$NH$_2$	17.6
At$^-$ + C$_6$H$_5$I + C$_4$H$_9$NH$_2$	17.2

Accordingly, the highest yields were those obtained using n-butylamine (143).

Samson and Aten (126, 127) were the first to use recoil astatination to synthesize astatobenzene, by irradiating (C$_6$H$_5$)$_3$Bi with α-particles accelerated in a synchrocyclotron. Astatobenzene, as well as the products of recoil astatine formed in the nuclear reaction, were separated and identified by GLC. Hot homolytic replacement reactions by recoil ^{211}At in situ have also been used to produce astatobenzene (111, 145, 150). Recoil astatine is produced from the electron capture decay of ^{211}Rn, which is one of the spallation products formed by bombardment of thorium or uranium with 660-MeV protons (16, 28, 85). After the separation of ^{211}Rn from other spallation products and its subsequent purification (82), carrier-free ^{211}Rn has been introduced into thoroughly evacuated glass ampuls filled with benzene or with the corresponding benzene derivative. The ampules were sealed and the ^{211}Rn ($\tau_{1/2}$ = 14.6 hours) was allowed to decay for 14–20 hours until the radioactive equilibrium with ^{211}At ($\tau_{1/2}$ = 7.2 hours) was attained. Organic and inorganic fractions were separated by extraction of the substrate with CCl$_4$ and aqueous NaOH solutions containing a small amount of Na$_2$SO$_3$ as a reducing agent. Identification and determination of yields of individual organic products were performed by GLC and HPLC. The highest yield was obtained with chlorobenzene as the substrate, when diluted with compounds of a lower IP than that of astatine, thus promoting neutralization of the originally multicharged recoil ^{211}At (145).

Most of the physical properties of astatobenzene have been determined (vide supra) by extrapolation from data derived for other halobenzenes. These extrapolations are based upon a comparison of their respective gas chromatographic behavior. Vasáros et al. (136) used a variety of stationary phases to establish the retention indexes (I_x)[4] for substituted benzene derivatives, including those of astatine. A linear correlation between I_x and T_b for monosubstituted benzenes has been established (17). The dissociation energy of the C—At bond in astatobenzene has been determined experimentally by measuring the kinetics of its thermal decomposition using a modified version of the toluene carrier gas technique (139, 151). The properties of astatobenzene established so far are listed in Table V.

b. Astatotoluenes. Preparation of the isomers of astatotoluene from the corresponding toluidines has been achieved through decomposition of their diazonium salts (99, 105, 163), but with relatively low radiochemical yields (10–20%). The toluidines were dissolved in HCl or H_2SO_4 and converted into the corresponding diazonium salts by addition of aqueous $NaNO_2$ at $-5°C$. Excess $NaNO_2$ was destroyed by the addition of urea, then At^- (in Na_2SO_3 solution) was added and the mixture slowly heated to between 50 and 80°C and then cooled. The products were extracted with diethyl ether and subsequently washed with NaOH solution and dried over $CaCl_2$. Analysis of the final product was carried out by GLC (99, 105) and by TLC (163). Under such experimental conditions $[OH^-] \gg [At^-]$ as the latter is present only in tracer amounts (10^{-15}–10^{-11} M). Consequently, the low radiochemical yields can be explained by the almost overwhelming competitive hydrolysis of the diazonium salts to the corresponding phenols.

In order to explain some of the peculiarities of the decomposition reaction and isomer distribution of the products (99, 105), a reaction mechanism has been postulated that involves complex formation between At^- and the diazonium ion, followed by electron transfer, leading to the release of nitrogen, while the phenyl radical recombines with astatine according to Eq. (12).

$$o\text{-}CH_3C_6H_4\text{—}\overset{+}{N}\!\!\equiv\!\!N + At^- \longrightarrow [o\text{-}CH_3C_6H_4\text{—}N\!\!=\!\!N^+]At^-$$
$$\downarrow e^-$$
$$o\text{-}CH_3C_6H_4At \longleftarrow o\text{-}CH_3C_6H_4\cdot + At\cdot + N_2 \quad (12)$$

[4] The retention time of the measured compound with that of a standard compound (usually an *n*-hydrocarbon) under the same conditions, the retention index I_x was calculated as:

$$I_x = 100\left[\frac{\ln(x) - \ln(n)}{\ln(n+1) - \ln(n)} + n\right]$$

where $t(x)$ is the retention time of component x, $t(n)$ is the retention atoms, similarly for $(n + 1)$. All measurements were made under the same conditions; $t(n) \leq t(x) \leq t(n+1)$.

The physicochemical properties of astatotoluene isomers have been estimated by extrapolation from those of the corresponding lighter halotoluenes based upon their GLC behavior (*99, 137, 138*), calculated directly from the GLC retention volumes (*137*), and determined experimentally by measurement of their thermal decomposition (*151*); these are given in Table V.

Homogeneous halogen exchange has been employed in order to prepare the isomers of $AtC_6H_4CF_3$ from the corresponding isomers of $ClC_6H_4CF_3$ in the presence of the *n*-butylamine (*149*). Radiochemical yields of 30, 45, and 36% for *ortho-*, *meta-*, and *para-*astatotrifluorotoluene, respectively, were significantly lower than yields observed for astatobenzene formation under similar conditions (*vide supra*). Gas–liquid chromatography was not only used to identify the products but also to estimate the physicochemical parameters ΔH_{vap} and T_b (*149*); the dissociation energy of the C—At bond was determined by measuring the kinetics of their thermal decomposition (*151*). These values are summarized in Table V.

c. *Astatohalobenzenes.* This group of organic astatine compounds has been the one most extensively studied (*17*); the isomers of AtC_6H_4X (X = F, Cl, Br, I) were obtained in essentially the same way as with astatobenzene (*vide supra*).

Whereas negligible radiochemical yields have been reported for the attempted electrophilic substitution of halobenzenes in homogeneous systems (*140*), astatination of fluorobenzene in heterogeneous mixtures with strong inorganic acids occurs at $\sim 90\%$ yield. This can be effected at 190°C over a 30-minute period; the *ortho*:*meta*:*para* (25:5:70) distribution of the AtC_6H_4F isomers reflected the electrophilic character of the reacting astatine. Under the same conditions, the distribution of *ortho*:*meta*:*para* AtC_6H_4Cl isomers was 30:20:50 from C_6H_5Cl; the yield was 72%. The corresponding yield sharply decreased for similar reactions with the isomers of AtC_6H_4Br and AtC_6H_4I (8.1 and 2.5%, respectively). The decreasing capability of halobenzenes to undergo electrophilic substitution has been correlated with the increasing atomic number of the bound halogen (*148*).

Hydrogen substitution in C_6H_5F, C_6H_5Cl, and C_6H_5Br by AtCl and AtBr is less efficient than the competing halogen-replacement reactions in these systems (*vide supra*); the total radiochemical yields are only a few percent and are poorly reproducible (*43, 99, 104*).

Specific astatohalobenzenes can be conveniently prepared by using heterogeneous halogen-exchange reactions, starting from the corresponding bromohalobenzene isomer, under the same conditions used for

TABLE V

EXTRAPOLATED PHYSICOCHEMICAL PROPERTIES OF ASTATOBENZENE AND OTHER ASTATINATED SUBSTITUTED AROMATIC MOLECULES[a,b,c,d]

Compound	Mean T_b (°C)	H_{vap} (kJ mol^{-1})	R_{C-At} (cm^3 mol^{-1})	μ_{C-At} (D)	D_{C-At} (kJ mol^{-1})	IP (eV)
C_6H_5At	222 ± 3 (87)	41.6 (115)	20.1 (138, 145)	1.60 (99)	205 (141)	8.8 (115)
	212 ± 3 (125, 128)	43.4 (137)		1.66 (138)	187.9 ± 21.3 (139)	
	217 (141)			1.53 (17)	180.7 ± 9.1 (151)	
	219 ± 3 (99)					
	216 ± 2 (137)					
	211 (115)					
o-At$C_6H_4CH_3$	237 ± 4 ⎫	46.3 ⎫	20.3 ⎫		180.7 ± 9.8 ⎫	
m-At$C_6H_4CH_3$	240 ± 4 ⎬(99, 137)	46.6 ⎬(137)	19.9 ⎬(138)	1.51 (17)	181.2 ± 8.6 ⎬(151)	
p-At$C_6H_4CH_3$	237 ± 4 ⎭	46.7 ⎭	19.9 ⎭		181.6 ± 9.5 ⎭	
o-At$C_6H_4CF_3$	485 ± 2 ⎫	44.5 ⎫			176.6 ± 8.8 ⎫	
m-At$C_6H_4CF_3$	478 ± 1 ⎬(149)	43.3 ⎬(149)			177 ± 8.6 ⎬(151)	
p-At$C_6H_4CF_3$	481 ± 1 ⎭	42.6 ⎭			176 ± 9.0 ⎭	
o-AtC_6H_4F	213 ± 2 ⎫	44.6 ⎫			179.5 ± 8.8 ⎫	
m-AtC_6H_4F	206 ± 2 ⎬(137)	43.4 ⎬(137)	20.3 (138)		179.9 + 9.0 ⎬(151)	
p-AtC_6H_4F	209 ± 2 ⎭	42.5 ⎭			179.9 ± 8.2 ⎭	

o-AtC$_6$H$_4$Cl	258 ± 2	50.8	
m-AtC$_6$H$_4$Cl	255 ± 3 *(137)*	49.0 *(137)*	
p-AtC$_6$H$_4$Cl	253 ± 2	47.5	
o-AtC$_6$H$_4$Br	303 ± 3		173.6 ± 8.6
m-AtC$_6$H$_4$Br	304 ± 3 *(99)*		175.3 ± 8.7 *(151)*
p-AtC$_6$H$_4$Br	305 ± 3		179.5 ± 9.1
o-AtC$_6$H$_4$I	336 ± 4		176.1 ± 7.9
m-AtC$_6$H$_4$I	337 ± 4 *(99)*	20.2 *(138)*	177.0 ± 8.6 *(151)*
p-AtC$_6$H$_4$I	337 ± 4		178.2 ± 9.3
o-AtC$_6$H$_4$NO$_2$	303		
m-AtC$_6$H$_4$NO$_2$	297 *(99)*		
p-AtC$_6$H$_4$NO$_2$	303		

[a] References in parentheses.
[b] GLC data *(87, 99, 125, 128)*.
[c] GLC-retention index (I_x) data *(137, 138)*.
[d] Thermal decomposition data *(139, 151)*.

preparing astatobenzene. The radiochemical yields for *meta*-, *para*- and for *ortho*-astatohalobenzenes have been found to be 60–70% and 40–50%, respectively (*141, 142*). Likewise, homogeneous nucleophilic substitution of the bromine atom has also been used to prepare astatohalobenzenes; no cross-isomerization has been observed in the course of these halogen-exchange reactions (*146*).

Astatohalobenzenes can be synthesized directly in ^{211}At recoil (Table III) experiments either via hydrogen replacement by recoil astatine in monohalobenzenes or by halogen replacement in dihalobenzenes. In the former, total yields range from 5 to 15%, with an almost statistical mixture of *ortho*-, *meta*, and *para*-astatohalobenzene products (*145, 150*). The production of AtC_6H_4F from the corresponding ClC_6H_4F isomers has been achieved with yields of 14% without noticeable isomerization of the products (19).

Astatohalobenzenes have also been prepared from the corresponding haloaniline isomers by decomposition of their diazonium salts under conditions similarly described for astatotoluenes (*vide supra*). Here again, relatively low radiochemical yields (10–26%) were obtained. Again, this has been attributed to the competing reaction of hydroxyl ions present in the aqueous solution in a much higher concentration than At^-, leading to the by-product formation of phenols (*99, 100, 105*).

All astatohalobenzenes were identified by GLC; their thermodynamic properties were estimated by extrapolative gas chromatographic data related to the corresponding dihalobenzene isomers (*17, 99, 137, 138*) (see Table V).

d. Astatophenols and Derivatives. Electrophilic astatination of phenol can be accomplished with AtCl or AtBr; the product is a mixture of *ortho*- and predominantly *para*-AtC_6H_4OH, with 20–30% radiochemical yields (*99, 104*). However, the ortho and para isomers have been prepared in much higher yield by astatination of the corresponding chloromercuri derivatives (*160, 166*). Phenol is easily mercurated in the para position by $Hg(CH_3COO)_2$, followed by reaction with NaCl (*47*). Addition of At^- in NaOH solution (containing SO_3^{2-}) to the chloromercuric derivative was followed by I_2 carrier in $CHCl_3$ and by KI_3. The mixture was stirred at room temperature for 30–40 minutes, and the HgI_2 precipitate was either filtered off or dissolved in excess KI solution. The astatinated products formed according to Eq. (13) are

$$C_6H_5OH \xrightarrow{Hg(CH_3COO)_2} p\text{-}HgC_6H_4OH \xrightarrow{NaCl} p\text{-}HgClC_6H_4OH \xrightarrow{At^-} p\text{-}AtC_6H_4OH$$

(13)

extracted from the mixture with CH_2Cl_2 and identified by TLC. Among the products, small amounts of astatoiodophenols (<5%) were also noted. These were presumably formed by astatination and subsequent iodination of dimercurated phenol. Astatine was observed to have a higher reactivity with chloromercuri derivatives than iodine in similar reaction systems. Furthermore, astatination, though in lower yields, was also possible without iodine carrier whereas radioiodine in tracer amounts failed to react with some aromatic substances (e.g., aniline, nitrobenzene). Reaction of astatine with chloromercuric compounds in the absence of an iodine carrier provides further indication that a radical mechanism for the astatination of these aromatic moieties may be implicated, particularly as this could be explained by the easy oxidation of At^- to At^0 at lower pH values (166).

meta-Astatophenol has been synthesized by the diazonium salt intermediary of *meta*-aminoaniline, according to the reaction

$$m\text{-}At\text{-}C_6H_4NH_2 \xrightarrow[-5°C]{NaNO_2} m\text{-}At\text{-}C_6H_4N_2^+ \xrightarrow[45°C]{H_2O} m\text{-}AtC_6H_4OH \qquad (14)$$

This preparative scheme leads to only 30% yield due to the side reactions between the *meta*-astatoaniline diazonium salt and astatophenol, which cannot be eliminated even by continuous extraction of the product with *n*-heptane (167). All the astatophenols synthesized to date have been identified by either HPLC (99, 104) or TLC (160, 166, 167). Their dissociation constants (K_a) have been established from extraction experiments by measuring the relative distribution of compounds between aqueous borax buffer solutions and *n*-heptane as a function of acidity. On the basis of these derived values, the Hammett σ-constants and hence the field (F) and resonance (R) effects have been estimated for these compounds (167) (see Table VI). The field effect for astatine was found to be considerably weaker than that for other halogens; the resonance effect was similar to that for iodine (162).

para-Astatoanisole has been prepared by nucleophilic astatination of the Tl^{III}-di(trifluoroacetic) acid derivative of anisole; the product (70–90% yield) was identified by TLC (162).

e. Astatoanilines and Derivatives. Although *ortho*- and *para*-$AtC_6H_4NH_2$ can be prepared by recoil astatination of aniline (99, 150) or by its electrophilic substitution via AtCl and AtBr (99, 104), these are relatively inefficient synthetic procedures and poor yields are obtained. Slightly better yields (10–15%) have been recorded for the reaction of At^- with the corresponding arsanilic acids at 60°C (167). A much more efficaceous synthetic route is via the appropriate chloromercuri derivatives (166, 167), as used initially for obtaining astatophenols (47, 160, 166). However, in this case, mercuration is necessarily performed

TABLE VI

Dissociation Constants (K_a) and Estimated Hammett
σ-Constants, Field (F), and Resonance (R) Effects for
Astatophenols[a]

Compound		pK_a	σ	F	R
AtC_6H_4OH	ortho	8.92 ± 0.03			
	meta	9.33 ± 0.03	0.26	0.28	−0.08
	para	9.53 ± 0.03	0.18		
$AtC_6H_4NH_2$	ortho	3.03 ± 0.03			
	meta	3.90 ± 0.03	0.24	0.25	−0.05
	para	4.04 ± 0.02	0.18		
AtC_6H_4COOH	ortho	2.71 ± 0.02			
	meta	3.77 ± 0.02			
	para	4.03 ± 0.02			

[a] Data from refs. *160* and *167*.

under more drastic conditions because of the lower reactivity of aniline compared with phenol. Single-pot reactions can be performed; mercuration can be achieved with $Hg(NO_3)_2$ or $Hg(ClO_4)_2$ in strong acid solutions at 60°C after 3–4 hours of stirring, following by addition of At^-. The astatoanilines have been obtained with radiochemical yields of approximately 80% and extracted from the reaction mixture with *n*-heptane. Small amounts of astatoiodoanilines (*166*) were formed (< 5%), again apparently from dimercurated intermediates (*vide supra*).

meta-Astatoaniline has been synthesized by reduction of *meta*-astatonitrobenzene (*vide infra*) with $SnCl_2$ at 60°C; a 90% yield has been reported (*167*).

The astatoanilines were identified by HPLC (*99, 104, 150*) and TLC (*160, 166, 167*). The pK_a values established by extraction experiments using *n*-heptane and citrate buffer solutions are given in Table VI, along with estimates of the Hammett σ-values, field, and resonance effects.

para-Astato-*N,N*-dimethylaniline has been synthesized by astatination of a chloromercuri derivative with a 65% radiochemical yield and identified by TLC (*166*).

f. Astatonitrobenzenes. Under conditions of halogen exchange similar to those described for synthesizing astatobenzene, the heterogeneous At for Br exchange has been employed in order to obtain all three isomers of astatonitrobenzene from the corresponding bromo compounds with yields on the order of 70%. Isotopic exchange was carried out at 50–60°C in order to avoid thermal decomposition of both the substrate and product. Identification was undertaken using GLC and HPLC, although partial decomposition of the product occurred (*147*). *meta*-Astatonitrobenzene was also obtained via its chloromercuric derivative, which was obtained by the mercuration of nitrobenzene using $HgClO_4$ in strong acidic solution (*81*). The final product was obtained with a 95% radiochemical yield and separated by extraction with CH_2Cl_2 and identified by TLC (*166, 167*). The T_b values of the astatonitrobenzene isomers were determined by extrapolation of GLC retention indexes (*147*) (see Table V).

g. *Astatobenzoic Acids and Derivatives.* All three isomers of astatobenzoic acid have been prepared through their respective diazonium intermediates using a procedure similar to that described for the astatotoluene isomers (*69, 71*). Yields of up to 90% have been reported (*153, 159, 163, 167*); TLC and column chromatography have been employed for their identification. *ortho*-Astatobenzoic acid has also been obtained in high yield (70–90%) by reaction of At^- with the Tl^{III} di(trifluoroacetate) derivative of benzoic acid (*162*). Heterogeneous isotopic exchange has also facilitated the rapid and high yield ($\sim 60\%$); the synthesis of *meta*-astatobenzoic acid, its methyl ether, and *meta*-astatohippuric acid from their bromo analogs was carried out *in vacuo* at 200–250°C (*134*). The time period for such syntheses did not exceed 1.5 hours and the final products, *meta*-astatobenzoic and its methyl ether and *meta*-astatohippuric acid, were identified by GLC and TLC, respectively (*134*). The pK_a values of astatobenzoic acids (Table VI), as estimated from extraction experiments using citrate buffer solutions and *n*-heptane, have also been used for identification purposes (*167*).

Astatobenzoic acid isomers have also been prepared by heterogeneous isotopic exchange by using the corresponding iodobenzoic acid. Halogen exchange has been effected under molten conditions, or with the substrate dissolved in a molten inert high-dielectric solvent, such as acetamide; yields of 60–70% have been obtained (*35*).

h. *Astatinated Aromatic Amino Acids and Proteins.* The most effective method for the incorporation of astatine into aromatic amino acid molecules is via their chloromercuri derivatives; phenylalanine, 4-methoxyphenylalanine, tyrosine, and 3-iodotyrosine have been astatinated in this manner (*164, 166*). As the hydrogen atoms in the benzene ring of 4-methoxyphenylalanine and phenylalanine are relatively deactivated, mercuration could be achieved only by prolonged stirring (5 hours) of suspensions of these substrates with $HgSO_4$ in 0.4 *N*

H_2SO_4 at 60°C. 4-Astatophenylalanine and 3-astato-4-methoxyphenylalanine were obtained by subsequent reaction of the nonisolated mercurated compounds with At^- over a period of about 30 minutes. The respective radiochemical yields were on the order of 85 and 70%; both astatoamino acids were identified by paper electrophoresis (*166*).

Mercuration of tyrosine and 3-iodotyrosine has been achieved smoothly at room temperature, and subsequent astatation (*vide supra*) led to 3-astatotyrosine and 3-astato-5-iodotyrosine, respectively. Yields were on the order of 60–80%; these compounds were identified by paper electrophoresis and paper chromatography (*166*). In these molecules, the precise position of the astatine atom in the aromatic ring has not been determined, but it was assumed on the basis of the electrophilic substitution pattern for mercuration. Synthesis of astatotyrosine by electrophilic astatation of L-tyrosine, using H_2O_2 as the oxidizing agent, has also been reported; yields were very low and the product appeared to be unstable under the experimental conditions described (*155*). Visser *et al.* (*164*) have discussed this problem thoroughly and their later investigations on the stability of astatinated amino acids in different media indicated that compounds such as 3-astatotyrosine and 3-astato-5-iodotyrosine, although stable in acid solutions, decomposed rapidly at pH ≥ 8, especially in the presence of oxidizing agents (*154, 161, 164*). Anecdotally, the synthesis of an astatoiodotyrosine, in which astatine was allowed to react with tyrosine in the presence of *N*-iodosuccinimide, has also been reported (*68, 70*).

In view of the potential therapeutic applications of ^{211}At (*vide infra*) the synthesis of stable astatinated protein molecules has attracted much effort (see Table VII). Proteins labeled with ^{211}At can be prepared most reliably and unambiguously via incorporation of previously prepared *para*-AtC_6H_4COOH by an acylation reaction with protein amino groups (*53, 156, 158, 159, 178*). Labeling proteins by this method was first reported by Hughes *et al.* (*69, 71*).

Acylation of protein amino groups by the mixed anhydride of *para*-AtC_6H_4COOH can be achieved by the following reaction sequence [Eqs. (15)–(17)].

$$p\text{-}AtC_6H_4COOH + N(C_4H_9)_3 \longrightarrow p\text{-}AtC_6H_4COO[HN(C_4H_9)_3] \qquad (15)$$

$$p\text{-}AtC_6H_4COO[HN(C_4H_9)_3] + C_2H_5COOCl \longrightarrow$$
$$p\text{-}AtC_6H_4COOCOC_2H_5 + [HN(C_4H_9)_3]Cl \qquad (16)$$

$$p\text{-}AtC_6H_4COOCOC_2H_5 + H_2N\text{-protein} \longrightarrow$$
$$C_2H_5OH + CO_2 + p\text{-}AtC_6H_4CONH\text{-protein} \qquad (17)$$

i-Butyl chlorocarbonate in 1,4-dioxane was added to a mixture of pure *para*-AtC$_6$H$_4$COOH in tetrahydrofuran containing N(C$_4$H$_9$)$_3$. This was allowed to stand for half an hour at room temperature; solvents were removed by vacuum pumping, then the protein in borate buffer (pH = 9) was added. Acylation was completed within 1 hour at 4°C; the ^{211}At-protein was separated from low-molecular-weight components by gel filtration, using sterile phosphate-buffered saline as eluent. Overall labeling yields are highly variable, and have ranged from 10 to 28% (*153, 157, 178*). Astatination under carrier-free conditions and using HPLC for the preparation of *p*-AtC$_6$H$_4$COOH (*63*) has resulted in a more consistent product yield ($\sim 30\%$).

Electrolytic generation of ^{211}At$^+$ for direct labeling of proteins (*1, 135*) and surface proteins on lymphocytes (*113*) has been best achieved at low electrode potentials (~ 1 V), thus avoiding the electrolytic denaturation of the ^{211}At-proteins (*1*). In bioorganic systems, the ^{211}At label has been lost rapidly (*113, 135*). A more reliable method for electrophilic astatination of proteins has involved using H$_2$O$_2$ as an oxidizing agent in a solution of At$^-$ containing KI as a carrier in neutral sodium phosphate buffer (*1*). Label yields of 60% were obtained for a synthesis period of 0.5–1.0 hour. Labeled proteins were separated by gel filtration.

Whereas electrophilic astatination of protein moieties has also been performed by others, the true nature of the At—protein bond has remained in doubt and, consequently, controversial views have been expressed about the mechanism of At-labeling. Initially it was assumed that At$^+$ was originally bound to protein tyrosine residues but was readily released due to the lability of the C—At bond. The free astatine as At^0 then reacted nonspecifically with other functional groups in the protein, eventually being trapped in the tertiary structure (*156*). However, it has been considered unlikely that electrophilic astatination of tyrosine occurs, as under similar experimental conditions decomposition of otherwise synthesized astatotyrosine has been observed in the presence of oxidizing agents (*161, 164*). Visser *et al*. (*169*) suggested two different types of labeling mechanisms. Relatively stable high radiochemical yields of astatinated proteins containing SH groups have been obtained even in the absence of oxidizing agents; formation of S—At bonds is implicated. In contrast, labeling of proteins with no free SH groups requires the presence of H$_2$O$_2$ as an oxidizing agent. In this latter case, the astatination process is slower and the yield and stability of the astato-labeled product are much lower. It has been assumed that the At^{3+} species HAtO$_2$[AtO(OH)] in the presence of H$_2$O$_2$ (pH = 7) forms a complex bond with oxygen and nitrogen atoms of

TABLE VII

ASTATINATION OF PROTEIN MOLECULES

Astatination method		Protein	Isolation	Yield (%)	Reference
Electrophilic (At⁺)	At/H$_2$O$_2$	Bovine serum albumin		80–90	169, 171
		β-Lactoglobulin		70–90	
		Hemoglobin		80–90	
		Cytochrome c	Gel filtration	1–35	169
		Lysozyme		1–25	
		Rat IgG and light chain fragment			156
	At/KI	Lymphocytes		~60	113
		Streptokinase			
		Phytohemoagglutin			
		Tuberculin (PPD)			
		Keyhole limpet Hemocyanine			
		Human γ-globulin	Gel filtration		
		Tuberculin			1

Acylation via p-AtC$_6$H$_4$COOH	Bovine serum albumin	Gel filtration	53, 69, 71, 178[a]
	γ-Globulin		71
	Fibrinogen		71
	Rabbit anti-mouse	Gel filtration	153
	Thymocyte IgG		12–28
	Concavalin A	Gel filtration	157, 159
	Human serum albumin		159
	BK 19.45 IgG		159
	BK 19.9 IgG		14, 158
	Rabbit IgG (polyclonal) (carrier free)	HPLC (p-AtC$_6$H$_4$COOH) Gel filtration	~30 63
At$^-$/KI	Ovalbumin		70
	γ-Globulin		70
At$^-$/N-I-succinimide	γ-Globulin		68, 70

[a] Acylation via p-AtC$_6$H$_4$SO$_2$Cl.

the proteins (*165*), similar to some oxometallic species. Such complex bonds are easily destroyed by reducing agents at pH > 8 and are not even stable enough to survive gel separation methods (*169*).

i. Astatonaphthoquinones. Astatination of 6-chloromercuri-2-methyl-1,4-naphthoquinone has been achieved in high yield (∼70%) by refluxing an ethanolic suspension of the substrate with At^-/ICl for 2 hours; the product was identified by TLC (*31*). A more rapid and efficaceous synthesis was achieved by *in vacuo* heterogeneous isotopic exchange with 6-iodo-2-methyl-1,4-naphthoquinone at its melting point (*35*). The preparation of 6-astato-2-methyl-1,4-naphthoquinol diphosphate (6-^{211}At-MNDP) has been accomplished by two methods. The first involves reduction, phosphorylation, and hydrolysis of previously synthesized 6-astato-2-methyl-1,4-naphthoquinone; overall, this five-step procedure took ∼7 hours, thereby reducing the original high radiochemical efficiency and resulting in only ∼15% yield (*31*). Its identification was determined by TLC (see Fig. 3). Alternatively, high-activity and specific activity 6-^{211}At-MNDP has been quickly synthesized by solid-phase thermal isotopic At^-/6-I-MNDP exchange *in vacuo* (*32*) (see Fig. 4). The presence of SO_3^{2-} markedly diminished the

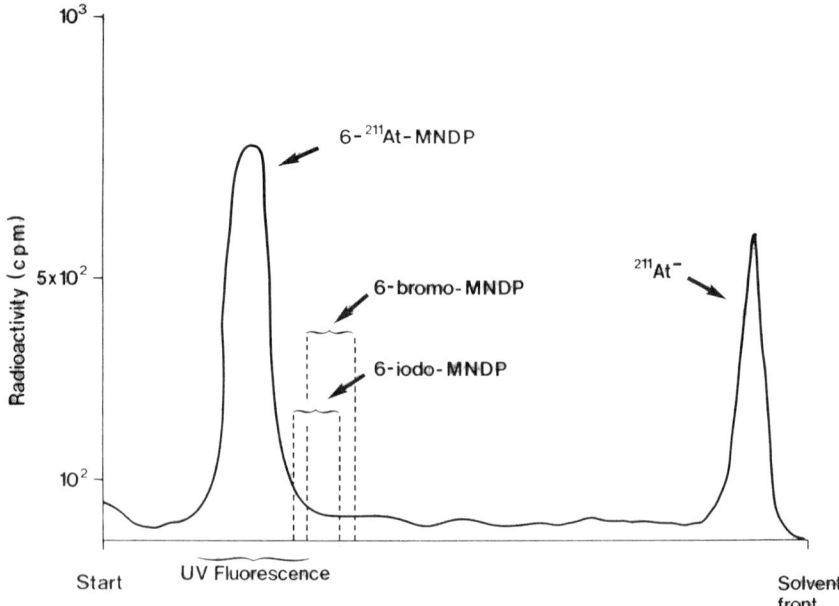

FIG. 3. Sequential TLC analysis [silica gel UV_{250} plates and n-BuOH/CH_3COOH/H_2O (10:7:3 v/v)]: identification of 6-^{211}At-MNDP (*32, 33*).

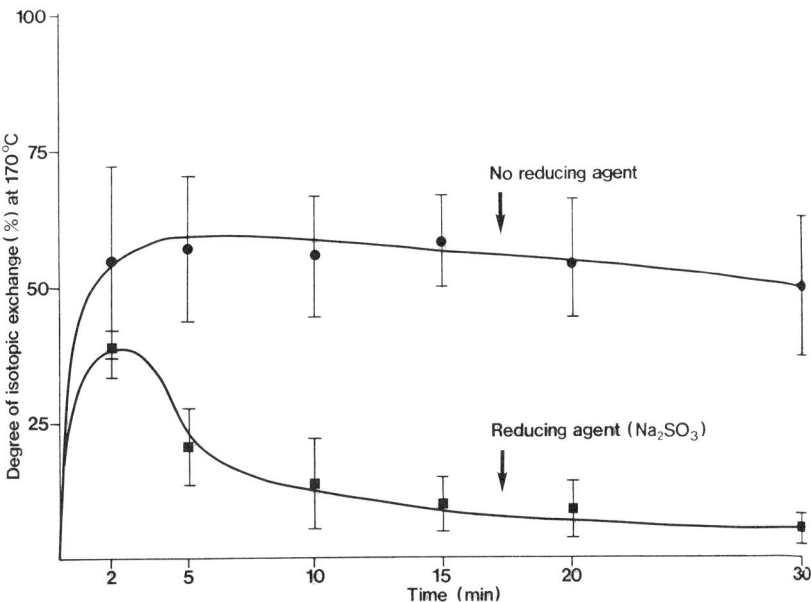

FIG. 4. Synthesis of 6-^{211}At-MNDP by *in vacuo* thermal heterogeneous isotopic exchange (^{211}At$^-$/I) (*32*).

product yield due to substrate decomposition. After purification by ion-exchange chromatography, radiochemical yields of 40–60% non-carrier-free 6-^{211}At-MNDP were obtained after a 10-minute reaction period. It has been postulated that the isotopic-exchange mechanism is facilitated through nucleophilic attack by At$^-$ at the 6-position of 6-I-MNDP, where resonance stabilization of the intermediary adduct would be additionally favored by protonation of the naphthalene nucleus (*33*).

j. Astatosteroids. Several complex steroid molecules have been astatinated via their chloromercuriderivatives; a mixture of 2-astatoestradiol (**I**), 4-astatoestradiol (**II**), and 2-astato-4-iodoestradiol (**III**) was obtained with yields of 55, 19, and 18%, respectively (*170*). Estradiol was mercurated by Hg(CH$_3$COO)$_2$ in a C$_2$H$_5$OH–water solution for 16 hours at room temperature, and then allowed to react with At$^-$ in H$_2$SO$_4$ in the presence of KI$_3$ for 1 hour. The products were separated and identified by TLC (*170*).

Generally speaking, astatoestradiols are less stable than the analogous iodine compounds. Their relative deastatination has been measured in mixtures with methanol and aqueous buffer solutions; no decomposition was observed at room temperature in acidic media up to

(I): R_1 = At; R_2 = H
(II): R_1 = H; R_2 = At
(III): R_1 = At; R_2 = I

(IV)

(V)

neutral pH. However, at 50°C and at pH 7, more than 75% of astatine was lost from the molecule over 20 hours. It was found that higher pH and the presence of H_2O_2 enhanced breakage of the C—At bond (170). Similarly, 6-astatocholesterol (IV) was also synthesized with 95% radiochemical yield and identified by TLC. However, heating the compound in ethanol–water solutions to 70°C and incubation at room temperature with H_2O_2 or with $NaHSO_3$ (170) did not produce deastatination. In vivo animal studies have also confirmed the nonlability of the 6-position C—At bond in the cholesterol molecule (171).

More recently, 6-astatomethyl-19-norcholest-5(10)-en-3β-ol (V) has been prepared rapidly, in high yield (60–70%), and at high specific activity by halogen exchange (At^-/I) in the presence of crown ethers. The crown ethers may fulfill a catalytic role by acting as specific cationic "sinks," and thus facilitating rapid nucleophilic exchange. The product was identified by TLC with an R_F value very close to that of the iodinated derivative (95).

3. Heterocyclic Compounds

a. Astatoimidazoles. Mercuration of imidazoles can be achieved over 3–5 hours by reaction with either $HgSO_4$ in H_2SO_4 solution at 55°C or with $Hg(CH_3COO)_2$ in aqueous $NaCH_3COO$ solution (pH = 5) at 50°C (46). Chloromercuri derivatives can be isolated, identified, and subsequently astatinated at room temperature using At^- with a molecular iodine carrier in sulfuric acid solution for 30 minutes. Using this method, 4-astatoimidazole, 5-astato-4-methylimidazole, and 5-astatohistidine have been prepared with yields ranging from 50 to 80%. These were identified by TLC and paper electrophoresis. Additionally, 2-astato-4-iodoimidazole and 2-astato-5-iodo-4-methylimidazole were also formed (~5%), presumably due to side reactions with the iodine carrier (168). These astatinated imidazoles were found to be stable in aqueous solutions at room temperature in the pH range 0–14, over a period of 15 hours. If the temperature was increased to 80°C, subsequent addition of Na_2SO_3 or H_2O_2 led to their rapid decomposition (*168*). It was also noted that excess iodine also gave rise to decomposition of 5-astatohistidine in a similar manner, as has been reported for the analogous iodine compound 5-iodohistidine (*130*).

b. Astatopyrimidines, Their Nucleosides, and Nucleotides. Preparation of 5-astatouracil was initially effected by astatination of the 5-aminouracil via its diazonium salt, in a manner similar to that described for obtaining astatohalobenzenes; the radiochemical yield was about 30% (*99, 102, 123*), and the product was identified by HPLC using the sequential analysis of uracil and other halouracils (*99, 102, 122*). Alternatively, 5-astatouracil can be synthesized from its chloromercuri derivative, with a much higher yield (~80%) and a very pure product (>95%) because of the absence of side reactions. Uracil can easily be mercurated with $HgSO_4$ in 0.4 N H_2SO_4 at room temperature for 3 hours; the chloromercuri compound can then be allowed to react with At^- according to the method employed for preparing astatophenol derivatives. The final product, 5-astatouracil, was isolated from the reaction mixture by extraction with benzene or *n*-butanol and identified by TLC (*166*). Attempts to produce astatouracil by allowing AtCl or AtBr to react with uracil or iodouracil have been unsuccessful (*99*).

In vitro studies have indicated that 5-astatouracil is chemically stable over the pH range 1–11.5 at room temperature, and 1–7 at 50°C over a period of 20 hours; these results are similar to those for the iodo analog. Heating to 50°C at pH > 11.5 resulted in loss of 20–30% of bound astatine after 20 hours. This has been attributed to direct attack of OH^- on the 5-position of the pyrimidine nucleus, as in the case of 5-iodouracil, Both halouracils are stable in the presence of SO_3^{2-} and H_2O_2 (*170*). From distribution studies utilizing benzene and aqueous borax buffers, the pK_a of 8.97 has been established for 5-astatouracil at 0°C (167).

In a similar manner, 5-astatouridine and its mono- and triphosphate derivatives (At-UMP and At-UTP) have been synthesized from the corresponding chloromercuri derivatives, which were formed first by the reaction of uridine, UMP, and UTP with $Hg(CH_3COO)_2$ in $NaCH_3COO$ buffer (pH 5) at 50°C over 3–5 hours (*170*). Astatination was accomplished in 1 hour with 75% radiochemical yield. 5-Astatouridine and 5-astato-UMP were purified and isolated by TLC; 5-astato-UTP was isolated by paper electrophoresis. Whereas the stability of 5-astatouridine was very similar to that of 5-astatouracil, the astatinated nucleotides were found to be sensitive to hydrolysis. Dephosphorylation of 5-astato-UTP and 5-astato-UMP was significant (50%) within 20 hours if acidic solutions (pH ~1) were kept at 50°C (*170*).

The nucleotide 5-astato-2-deoxyuridine has been obtained from 2-deoxyuridine in a manner similar to that described for preparing 5-astatouridine with a radiochemical yield of ~85% (*170*). Attempts to synthesize 5-astato-2-deoxyuridine from the diazonium salt of 5-amino-2-deoxyuridine led only to a 2–3% yield of desired product, whereas 20–25% of the bound astatine was found in the form of 5-astatouracil (*99, 123*). This was apparently due to hydrolysis of the *N*-glycosyl bond in the course of the diazotization reaction. The final product, 5-astato-2-deoxyuridine, was identified by TLC, paper electrophoresis (*170*), and HPLC (*99, 122, 123*).

Other nucleotides, such as 5-astatocytosine, 5-astatocytidine, as well as the monophosphate derivatives of 5-astatocytidine (At-CMP) and of 5-astato-2-deoxycytidine (At-dCMP), have been similarly prepared via astatination of their chloromercuri derivatives (*170*). Product yields are of a similar order; final compounds were isolated and identified by TLC and by paper electrophoresis. It was found that 5-astatocytosine and 5-astatocytidine were stable at room temperature over a wide pH range. Their behavior in the presence of reducing or oxidizing agents paralleled that of their iodo analogs; deastatination occurred. In acidic solutions (pH = 1), dephosphorylation of 5-astato-CMP has been observed in addition to the slow decomposition of the sugar–pyrimidine bond. However, these effects were pronounced for 5-astato-dCMP which was found to decompose completely within 20 hours at 50°C in acid solutions, followed by the formation of 5-astatocytosine and 5-astato-2-deoxycytidine.

Variously astatinated nucleic acids, At-DNA and At-RNA, have been obtained via their chloromercuri derivatives with radiochemical yields of >90%. These compounds have been isolated and proved stable to purification by gel filtration. There was no evidence of any deastatination at pH 2–11 on incubation for 20 hours, nor at neutral pH in the presence of small amounts of reductants or oxidants at room temperature. However, heating to 50°C caused slow deastatination with 15–20% astatine loss in 20 hours. On heating of the At-nucleic acids there was some degree of degradation but this did not appear to involve breakage of the C—At bond (*170*).

c. *Astatophenazathioniums.* Syntheses of astatinated derivatives of 3,7-bis(dimethylamino)phenazathionium chloride (methylene blue) have been attempted by several synthetic routes (*41, 94*).

(VI): R_1, R_2 = H; R_4 = ^{211}At
(VII): R_1 = I; R_2 = ^{211}At; R_3 = H

Astatination of the diazonium salt of 4-amino-methylene blue led to only ~10% yield of 4-astato-methylene blue (**VI**). Alternatively, 4-astato-methylene blue was rapidly synthesized by heterogeneous nucleophilic isotopic exchange in the presence of 18-crown-6 ether at 80°C, with 4-iodo-methylene blue. The product was separated and identified by TLC; yields ranged from 50 to 65% (*41*). Additionally, 2-astato-8-iodo-methylene blue (**VII**) has similarly been prepared in 60% yield from 2,8-diiodo-methylene blue (*41*).

Attempts at electrophilic astatination of methylene blue with ^{211}At/chloramine-T have proved ineffective (*41*).

IV. Biological Behavior

Astatine, regardless of its electronic state (*vide infra*), appears to possess many physiobiochemical properties similar to those of its nearest homolog, iodine (*34, 57, 60, 119, 176*). However, astatine also exhibits a proclivity to accumulate in macrophage-laden tissue, such as lungs, liver, and spleen (*37, 60, 119*); this has been attributed to its amphoteric character.

In rats, guinea pigs, monkeys, and man uptake of [^{211}At]-astatide anion into thyroid tissue has been demonstrated as high, but to a lesser degree than that for radioactive iodine (*57, 59, 60, 132*). Unlike radioactive iodine, the tissue:plasma ratios for [^{211}At]-astatide anion have all been found generally greater than unity (*60*). Its uptake into human thyroid tissue is relatively greater (*59*). Small doses of ^{211}At have been observed to greatly modify thyroid function in animals, with no apparent damage to important contiguous functional anatomical structures such as parathyroid tissue (*58*). Like radioactive iodine, uptake of ^{211}At into the thyroid gland can be blocked by the prior administration of iodide, thiocyanate (*60, 133*), or perchlorate ions (*33, 37*). Astatine can be leached from thyroid tissue by thiocyanate, (*133*) presumably due to the formation of stable At–SCN complexes. Paradoxically, other agents that interfere with the organic binding of iodine in thyroid tissue, such as thiouracil or propylthiouracil, cause increased uptake of ^{211}At into thyroid tissue (59, 132). The possible biochemical fate of *in vivo* ^{211}At can be schematically summarized as shown in Scheme 3. If ^{211}At is administered as a radiocolloid, it has been found to localize in liver (60) and lungs (33); this behavior is common to most colloids *in vivo* (33).

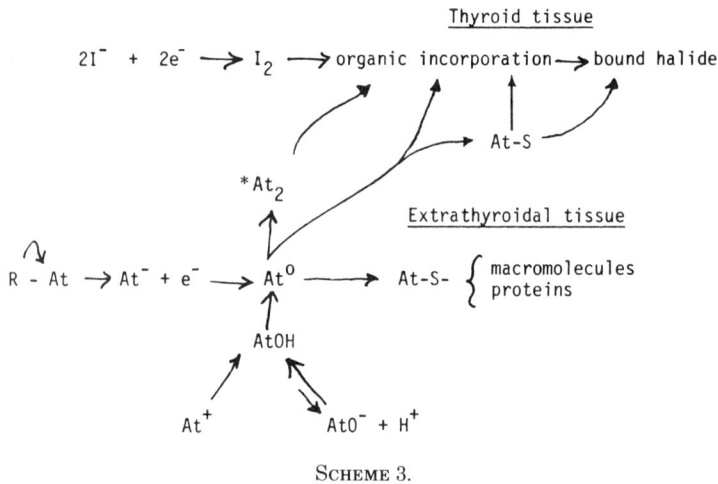

SCHEME 3.

It has been demonstrated that ^{211}At is embryotoxic in pregnant mice, and there also exists a dose-related occurrence of associated fetal malformations (29, 30). Long-term studies in female rats that had received 0.5 μCi g^{-1} ^{211}At$^-$ systemically indicated a significant incidence of radiation-induced mammary carcinomata (39/45; 44%) and endometrial polyps (43/55; 76.4%) at 14 months after treatment. No thyroid tumors were found (50). Detailed macroscopic radiation dosimetric studies related to the biodistribution of ^{211}At$^-$ in animal models have been reported (29, 33, 39).

V. Biomedical Applications—Cancer Therapy

Astatine-211 ($\tau_{1/2} = 7.21$ hours) possesses many of the desired physical (Fig. 5), chemical, and radiobiological properties thought pertinent to its possible application in cancer therapy (26, 33, 34, 36, 40). Astatine-211 decays along two branches: (1) by direct α-particle decay (41.94 $\pm$ 0.50%; 5.87 MeV) to ^{207}Bi ($\tau_{1/2} = 38$ years), which decays by electron

---
* Formation of At$_2$ is statistically highly unlikely.

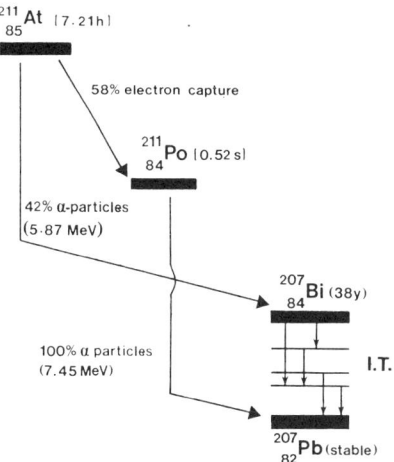

FIG. 5. Decay scheme for ^{211}At.

capture to ^{207}Pb; and (2) by electron capture to ^{211}Po ($\tau_{1/2} = 0.52$ second), which is in transient equilibrium with ^{211}At and which subsequently decays by the emission of α-particles almost entirely of energy 7.45 MeV to form stable ^{207}Pb. There is one α-particle per disintegration. The ranges of the α-particles of ^{211}At in unit density tissue are either 55 μm, corresponding to energy 5.87 MeV for 42% of the disintegrations, or 80 μm, corresponding to energy 7.45 MeV for approximately 58% of the disintegrations. The total mean absorbed radiation dose in tissue is 150.6 cGy g μCi^{-1} (119). The contribution from the long-lived decay of ^{207}Bi is negligible (<0.002%).

The mean dose-average linear energy transfer (LET$_\infty$) of ^{211}At-emitted α-particles is approximately 98.84 keV μm^{-1} unit density tissue (73, 175). This value is probably very close to the optimum for endoradiotherapeutic effects (56). In vitro cell studies have demonstrated that the cell-killing effect of such high-LET radiation is independent of dose rate. As a result of its intense focal ionizing nature, cell damage is lethal, being predominantly attributed to nonrejoining double strand breaks in DNA (121). There has been no evidence of repair of sublethal damage or potentially lethal cellular damage (9, 10, 21).

Conclusions drawn from in vitro cell studies with heavy ion beams of varying LET$_\infty$ have indicated that approximately 100 keV μm^{-1} was optimal in achieving a maximum relative biological effect (RBE), as demonstrated by the blocking of cells in $G_2 + M$ and subsequent lethal effects (21, 96, 97). The efficient arrest of cultured human squamous cell carcinoma of the larynx (HEp2) and murine rectal adenocarcinoma

TABLE VIII
Therapeutic Studies with Astatine and Its Labeled Compounds

Compound	Tumor	Species	Experimental details	Reference
^{211}At$^-$	Papillary thyroid adenocarcinoma	Man	Anecdotal study in a patient with cervical node metastases. No significant uptake of At into an excised node. Poorly differentiated tumor	59
^{211}At$^-$	Rectal adenocarcinoma	Mouse	Biodistribution studies in thyroid-blocked (ClO_4^-) animals. High uptake in spleen, lungs, and stomach. No therapeutic effect over 1- to 20-μCi dose range	33, 37–39, 109
^{211}At$^-$, At^0, At$^+$	Sarcoma 180	Mouse	Biological fate was identical regardless of administered valence state. No tumor affinity; high uptake into spleen, lung, liver, stomach, and thyroid	119
^{211}At–tellurium colloid	Ovarian adenocarcinoma	Mouse	Therapeutic studies with 25–200 μCi ^{211}At–Te colloid. Injected by an intraperitoneal route. Ascitic tumor. 100% cures in 25- to 50-μCi dose range; acceptable morbidity. Greater than 50 μCi radiation-induced lethality observed	25, 27
^{211}At$^-$	Benign epithelial cyst	Man	Trial local instillation into anterior segment of the eye. Retarded reaccumulation of cystic fluid, but cyst not destroyed, with such a modest dose (10 μCi)	131

Compound	Tumor/Target	Animal	Observations	Refs
5-^{211}At–uracil, 5-^{211}At–UdR	Sarcoma 180	Mouse	Rapid deastatination of compounds. Biodistribution profile similar to that for ^{211}At$^-$. No preferential tumor uptake, although three times that for the analogous iodo compound	99, 124
6-^{211}At–MNDP	Rectal adenocarcinoma	Mouse	Biodistribution studies: significant tumor uptake over 12 hours. Heterogeneous distribution within tumor tissue, related to alkaline phosphatase positive areas. Therapy with 0.125–20 μCi. Survival (Φ)–dose (D) relationship of a Langmuir-type saturation equation $\Phi_n = \zeta_n D(1 + \zeta_n D)^{-1}$ at n months. At 12 months, 50% survival plateau at $D > 2.5$ μCi. Permanent healing	33, 34, 37–40, 106–109
^{211}At–BK 19.9 monoclonal antibody	Human promyelocytic leukemia (HL60)	Nude mouse	Significant localization of ^{211}At in tumor by 12 hours postinjection. Generally, a higher uptake into spleen, lungs, and thyroid, suggestive of metabolic deastatination or free p-AtC$_6$H$_4$OH impurities	13, 158, 159
^{211}At–anti–thy 1,1 (OX7) monoclonal antibody	T-cell lymphoma	Mouse	Therapeutic studies. 78% and 47% of mice that received a single i.v. injection of 2.2–2.4 μCi, 48 h after injection of lymphoma cells survived >200 days (controls ^{211}At$^-$ survived only ~25 days)	63, 65
4-^{211}At–methylene blue	B16 melanoma	Mouse	High specific activity compound; marked therapeutic efficacy by lung colony assay. Localization of 4-^{211}At–methylene blue in intracellular melanin. No cytotoxic effect with ^{211}At$^-$	94

(CMT-93) cells in $G_2 + M$ after exposure to small doses of $^{211}At^-$ has been quantitatively confirmed by flow cytofluorometric methods (35).

In vitro clonal tumor cell studies have demonstrated the severe cytotoxicity of α-particles delivered by $^{211}At^-$; single exponential cell survival–dose curves were obtained, with D_0 (37% survival) values of 29–48 cGy and 57–73 cGy for Chinese hamster V-79 (61) and HEp2 cells, respectively (33). In both studies the oxygen enhancement ratio (OER) was found to be slightly greater than unity, probably resulting from the low-LET components of ^{211}At decay (see Fig. 5). In biological systems, such α-particle emissions enable comparable cytotoxicities to be effected in both hypoxic and euoxic tumor cell populations.

The results from similar clonogenic survival studies with an [^{211}At]-IgG monoclonal antibody to human leukemia cells (158) have given a mean D_0 of 12 ^{211}At atoms cell^{-1}; the RBE has been determined as approximately 4, when compared to the γ-radiations from cobalt-60 (13). A range of RBE values (2.8–5.2) for ^{211}At α-particles compared with other low-LET (γ, β^-) radiations has been obtained for a variety of tissues under different in vitro and in vivo experimental conditions (12, 13, 29, 61, 64).

Apart from isolated studies of the selective suppression of graft-versus-host mechanisms by ^{211}At-labeled proteins in organ transplantation (113), and various in vitro immunological investigations (1, 135), the main in vivo applications of ^{211}At has been directed toward tumor therapy (25, 27, 33, 38–40, 65, 108, 109). Astatine-211 can be stably incorporated into potentially useful biologically active organic compounds via covalent ^{211}At—C bonds, both rapidly and with high yield and specific activity (32, 33, 160). In several complex ^{211}At-biomolecules, the ^{211}At—C bond has been found to be metabolically stable in vivo (37, 63, 160, 171). Applications of such compounds and ionic astatine as potential endoradiotherapeutic drugs have been widely studied in both animal and human tumor models (see Table VIII).

It has been shown that [^{211}At]-endoradiotherapy can be curative in tumor bearing animals, without untoward acute or chronic side effects (27, 33, 39, 65). On the basis of results derived from localized ^{211}At–Te colloid therapy (25, 27), such an approach might be advocated for the treatment of malignant intraperitoneal disease, and perhaps chronic synovitis, in human subjects (27). From the point of view of the systemic application of ^{211}At, encouraging results have been derived from investigations with tumor-surface antigen-specific ^{211}At-labeled monoclonal antibodies (65). Likewise, the concurrent development and investigation of a novel class of metabolically directed radiohalogenated potential anti-cancer drugs have proved most promising. Of these,

6-^{211}At-MNDP (*31–33, 36, 108*), whose mechanism of intracellular localization is related to the presence of oncogenically expressed tumor-membrane alkaline phosphatase isoenzymes (*42, 108*), has been demonstrated strikingly effective in an animal tumor model (*33, 34, 38, 39*). It has also served as a concomitant analytical probe for identifying the intracellular locus of radiotherapeutic action of this class of drug by α-particle track autoradiography (*33, 106–109*). Phase I and II human therapeutic trials are shortly envisaged (*33, 34*).

Clearly, there is much scope for the future development of ^{211}At-labeled molecules, and for investigation of their possible role in cancer therapy (*33*).

NOTE ADDED IN PROOF. The facile astatination of *para*-AtC$_6$H$_4$COOH and 3-At-tamoxifen by astatination of 4-(*n*-C$_4$H$_9$)$_3$Sn-benzoic acid oxazoline and 3-(*n*-C$_4$H$_9$)$_3$Sn-tamoxifen via an electrophilic destannylation route under mild conditions has recently been reported (*179*). Product isolation and identification were achieved HPLC and TLC; radiochemical yields were 80% and 60%, respectively. Attempted astatination under carrier(I$_2$)-free conditions gave negligible yields (0.5–1%). Results suggested that the electrophilic species was either AtI or AtI$_2$ as analogous iodo-destannylation reaction kinetics exhibit a second-order dependence upon iodide concentration (*180*).

ACKNOWLEDGMENTS

I am most grateful to Professor J. S. Mitchell, FRS, for his continued advice and encouragement. Also, many thanks to Dr. Klara Berei, with whom I recently had useful discussions. Mary-Ann Starkey's valuable contribution to the preparation of the final manuscript is much appreciated.

REFERENCES

1. Aaij, C., Tschroots, W. R. M. J., Lindner L., and Feltkamp, T. E. W., *Int. J. Appl. Radiat. Isot.* **26**, 25 (1975).
2. Appelman, E. H., NAS-NS 3012, 1960.
3. Appelman, E. H., Ph.D. Thesis (UCRL-9025), University of California, Berkeley, 1960.
4. Appelman, E. H., Sloth, E. N., and Studier, M. N., *Inorg. Chem.* **5**, 766 (1966).
5. Appelman, E. H., *Int. Rev. Sci. Inorg. Chem. Ser. I* **3**, 181 (1972).
6. Aten Jnr., A. H. W., *Adv. Inorg. Chem. Radiochem.* **6**, 207 (1964).
7. Aten Jnr., A. H. W., Doorgeest, T., Hollstein, U., and Moeken, P. H., *Analyst (London)* **77**, 774 (1952).
8. Bächmann, K., and Hoffmann, P., *Radiochim. Acta* **15**, 153 (1971).
9. Barendsen, G. W., *Int. J. Radiat. Biol.* **8**, 453 (1964).
10. Barendsen, G. W., Broerse, J. J., and Breur K. (Eds.), "High-LET Radiations in Clinical Radiotherapy." Pergamon, Oxford, 1979.
11. Barton, G., Ghiorso, A., and Perlman, I., *Phys. Rev.* **82**, 13–19 (1951).
12. Basson, J. K., and Shellabarger, C. J., *Radiat. Res.* **5**, 502 (1958).

13. Bateman, W. J., Ph.D. Thesis, University of Birmingham, United Kingdom, 1983.
14. Bateman, W. J., Vaughan, A. T. M., and Brown, G., *Int. J. Nucl. Med. Biol.* **10,** 241 (1983).
15. Bell, R. P., and Gelles, E., *J. Chem. Soc.*, p. 2734 (1951).
16. Belyaev, B. N., Wang Yung-Yu, Sinotova, E. N., Nemeth, L., and Khalkin, V. A., *Radiokhimiya* **2,** 603–613 (1960).
17. Berei, K., and Vasáros, L., "The Chemistry of Functional Groups," (S. Patai and Z. Rappoport eds.), Suppl. D, Pt. 1, p. 405. Wiley, New York, 1983.
18. Berei, K., and Vasáros, L., "Astatine," Gmelin Handbook, 1985.
19. Berei, K., Vasáros, L., Norseev, Yu. V., and Khalkin, V. A., *Radiochem. Radioanal. Lett.* **26,** 177 (1976).
20. Bertholet, A., *C. R. Hebd. Seances Acad Sci.* **227,** 829 (1948).
21. Bertsche, U., Iliakis, G., and Kraft, G., *Radiat. Res.* **95,** 57 (1983).
22. Beyer, G. J., Dreyer, R., Odrich, H., and Roesch, F., *Radiochem. Radioanal. Lett.* **47,** 63 (1981).
23. Bimbot, R., Gardes, D., and Rivet, M. F., *Phys. Rev.* C **4,** 2180 (1971).
24. Bimbot, R., and Rivet, M. F., *Phys. Rev.* C **8:** 375 (1973).
25. Bloomer, W. D., McLaughlin, W. H., Neirinckx, R. D., Adelstein, S. J., Gordon, P. R., Ruth, T. J., and Wolf, A. P. *Science* **212,** 340 (1981).
26. Bloomer, W. D., and Adelstein, S. J., *Radioakt. Isot. Klin. Forsch., Badgastein, Wien* **15,** 227 (1982).
27. Bloomer, W. D., McLaughlin, W. H., Lambrecht, R. M., Atcher, R. W., Mirzadeh, S., Madara, J. L., Milius, R. A., Zalutsky, M. R., Adelstein, S. J., and Wolf A. P., *Int. J. Radiat. Oncol. Biol. Phys.* **10,** 341 (1985).
28. Bochvarova, M., Tyung, D. K., Dudova, I., Nerseev Yu. V., and Khalkin, V. A., *Radiokhimiya* **14,** 858 (1972).
29. Borrás, C., D. Sc. Thesis, Universidad Barcelona, Spain, 1974.
30. Borrás, C., Gorson, R. O., and Brent R. L., *Phys. Med. Biol.* **22,** 118 (1977).
31. Brown, I., *Int. J. Appl. Radiat. Isot.* **33,** 75 (1982).
32. Brown, I., *Radiochem. Radioanal. Lett.* **53,** 343 (1982).
33. Brown, I., M. D. Thesis, University of Cambridge, U. K., 1986.
34. Brown, I., *Appl. Radiat. Isotopes Part A* **37,** 789 (1986).
35. Brown, I., Unpublished results.
36. Brown, I., Carpenter, R. N., and Mitchell, J. S., *Eur. J. Nucl. Med.* **7,** 115 (1982).
37. Brown, I., Carpenter, R. N., and Mitchell, J. S., *Int. J. Appl. Radiat. Isot.* **35,** 843 (1984).
38. Brown, I., Carpenter, R. N., and Mitchell, J. S., *Int. J. Radiat. Biol.* **45,** 457 (1984).
39. Brown, I., Carpenter, R. N., and Mitchell, J. S., *Int. J. Radiat. Oncol. Biol. Phys.*, submitted.
40. Brown, I., "Nuclear Medicine and Biology" (Ed. C. Raynaud, ed.), Vol. I. Pergamon, Oxford, 1982.
41. Brown, I., Carpenter, R. N., Link, E., and Mitchell, J. S., *J. Radioanal. Nucl. Chem.*, in press (1986); *ibid, J. Lab. Comp. Radiopharm.*, in press (1986).
42. Carpenter, R. N., Brown, I., and Mitchell, J. S., *Int. J. Radiat. Oncol. Biol. Phys.* **9,** 55 (1983).
43. Cavallero, A., Ph.D. Thesis, Université Catholique, Louvain, Belgium, 1981.
44. Corson, D. R., Mackenzie, K. R., and Segrè, E., *Phys. Rev.* **58,** 672 (1940).
45. Corson, D. R., Mackenzie, K. R., and Segrè, E., *Nature (London)* **159,** 24 (1947).
46. Dale, R. M. K., Livingston, D. C., and Ward, D. C., *Proc. Natl. Acad. Sci. U.S. A.* **70,** 2238 (1973).
47. Dimroth, O., *Chem. Ber.* **31,** 2154 (1898).

48. Doberenz, V., Nhan, D. D., Dreyer, R., Milanov, M., Norseev, Yu, V., and Khalkin, V. A., *Radiochem. Radioanal. Lett.* **52,** 119 (1982).
49. Downs, A. J., and Adams, C. J., "The Chemistry of Chlorine, Bromine, Iodine and Astatine." Pergamon, Oxford, 1975.
50. Durbin, P. W., Asling, C. W., Johnson, M. E., Parrott, M. W., Jeung, N., Williamson, M. H., and Hamilton, J. G., *Radiat. Res.* **9,** 378 (1958).
51. Foreman, Jnr., B. M., and Seaborg, G. T., *J. Inorg. Nucl. Chem.* **7,** 305 (1958).
52. Freiesleben, H., Britt. H. C., Birkelund, J., and Huizenga, J. R., *Phys. Rev.* **C 10,** 245 (1974).
53. Friedman, A. M., et al., *Int. J. Nucl. Med. Biol.* **4,** 219 (1977).
54. Gesheva, M. Kolachkovsky, A., and Norseev, Yu, V., *J. Chromatogr.* **60,** 414 (1971).
55. Hahn, O., *Naturwissenschaflen* **14,** 758 (1926).
56. Hall, E. J., "Radiobiology for the Radiologist," Harper, New York, 1978.
57. Hamilton, J. G., and Soley, M. H., *Proc. Natl. Acad. Sci. U.S.A.* **26,** 483 (1940).
58. Hamilton, J. G., Durbin, P. W., and Parrott, M. W., *J. Clin. Endocrinol. Metab.* **14,** 1161 (1954).
59. Hamilton, J. G., Durbin, P. W., and Parrott, M. W., *Proc. Soc. Exp. Biol. Med.* **86,** 366 (1954).
60. Hamilton, J. G., Asling, C. W., Garrison, W. M., and Scott, K. G., *Univ. Calif. Berkeley Publ. Pharmacol.* **2,** 283 (1953).
61. Harris, C. R., Adelstein, S. J., Ruth, T. J., and Wolf, A. P., *Radiat. Res.* **74,** 590 (1978); Kassis, A. I., Harris, C. R. Adelstein, S. J. Lambrecht, R., and Wolf, A. P., *Radiat. Res.* **105,** 27 (1986).
62. Harris, R. G., Ph.D. Thesis, University of Birmingham, U. K., 1960.
63. Harrison, A., and Royle, L., *Int. J. Appl. Radiat. Isot.* **11,** 1005 (1984).
64. Harrison, A., and Royle, L., *Health Phys.* **46,** 377 (1984).
65. Harrison, A., and Royle, L., *Cancer Drug, Deliv.* **2,** 227 (1985).
66. Hoffmann, P., *Radiochim. Acta* **17,** 169 (1972).
67. Hoffmann, P., *Radiochim. Acta* **19,** 69 (1973).
68. Hughes, W. L., and Gitlin, D., Brookhaven National Laboratory, Upton, N. Y., BNL-314, 48 (1954).
69. Hughes, W. L., and Klinenberg, J., Brookhaven National Laboratory, Upton, N. Y., BNL-367, 42 (1955).
70. Hughes, W. L., and Gitlin, D., *Fed. Proc. Fed. Am. Soc. Exp. Biol.* **14,** 229 (1955).
71. Hughes, W. L., Smith, E., and Klinenberg, J., BNL-406, 44 (1956).
72. Hyde, E. K., and Giorso, A., *Phys. Rev.* **90,** 267 (1953).
73. Jardine, J., *Phys. Rev.* **C 11,** 1385 (1975).
74. Johnson, G. L., Leininger, R. F., and Segre, E., *J. Chem. Phys.* **17,** 1 (1949).
75. Karlik, B., *Acta Phys. Austriaca* **2,** 182 (1948).
76. Karlik, B., and Bernert, T., *Naturwissenschaften* **32,** 44 (1943).
77. Karlik, B., and Bernert, T., *Z. Phys.* **123,** 15 (1944).
78. Kelley, E. L., and Segre, E., *Phys. Rev.* **75,** 999 (1949).
79. Khalkin, V. A., and Herrmann, E., *Isotopenpraxis* **11,** 333 (1975).
80. Khalkin, V. A., Herrmann, E., Norseev, Yu. V., and Dryer, I., *Chem. Ztg.* **101,** 470 (1977).
81. Klapproth, W. J., and Westheimer, F. H., *J. Am. Chem. Soc.* **72,** 4461 (1950).
82. Kolachkovsky, A., and Norseev, Yu. V., Joint Institute for Nuclear Research, Dubra, U.S.S.R., JINR-P6-6923, 1 (1969).
83. Kolachkovsky, A., and Norseev, Yu. V., *J. Chromatog.* **84,** 175 (1973).
84. Kolachkovsky, A., and Khalkin, V. A., Joint Institute for Nuclear Research, Dubra, U.S.S.R., JINR-12-9473, 1 (1976).

85. Kurchatov, B. V., Mekhedov, V. N., Chistyakov, L. V., Kurnetsova, M. Y., Borrisiva, N. I., and Solovyev, V. G., *Zh. Eksp. Teor. Fiz.* **35,** 56 (1958).
86. Kuzin, V. I., Ph. D. Thesis, Leningrad State University, U.S.S.R., 1971.
87. Kuzin, V. I., Nefedov, V. D., Norseev, Yu. V., Toropova, M.A., and Khalkin, V. A., Soviet Radiochem. (*Engl. Transl.*) **12,** 385 (1970).
88. Kuzin, V. I., Nefedov, V. D., Norseev, Yu. V., Toropova, A., Filatov, E. S., and Khalkin, V. A., *High-Energy Chem.* (*Engl. Transl.*) **6,** 161 (1972).
89. Lambrecht, R. M., and Mirzadeh, S., *Int. J. Appl. Radiat. Isot.* **36,** 443 (1985).
90. Lavruhkhina, A. K., and Pozdnyakov, A. A., "Analytical Chemistry of Techetium, Promethium, Astatine and Francium." Humphrey, Ann Arbor, 1970.
91. Lefort, M., Simonoff, G., and Tarrago, X., *Acad. Sci.*, p. 216 (1959).
92. Lefort, M., Simonoff, G., and Tarrago, X., *Bull. Soc. Chim. Fr.*, p. 1726 (1969).
93. Lindner, L., Brinkman, G. A., Sver, T. H. G. A., Schimnel, A., Veenboer, J. Th., Karten, F. H. S., Visser, J., and Leurs, C. J., *IAEA, Vienna* **1,** 303 (1973).
94. Link, E., Brown, I., Carpenter, R. N., and Mitchell, J. S., *Proc. 6th Eur. Workshop Melanin Pigment. 6th, Murcia, Spain*, p. 33 (1985).
95. Liu, B.-L., Jui, Y.-T., Liu, Z.-H., Luo, C., Kojima, M., and Maeda, M., *Int. J. Appl. Radiat. Isot.* **36,** 561 (1985).
96. Lücke-Huhle, C., *Radiat. Res.* **89,** 298 (1982).
97. Lücke-Huhle, C., Blakely E. A., Chang P. Y., and Tobias C. A., *Radiat. Res.* **79,** 97 (1979).
98. Merinis, J., and Bouissieres, G., *Radiochim. Acta* **12,** 140 (1968).
99. Meyer, G.-J., Ph.D. Thesis (JUEL-1418), Universität Köln, Federal Republic of Germany, 1977.
100. Meyer, G.-J., Rössler, K., and Stöcklin, G., *Radiochem. Radioanal. Lett.* **21,** 247 (1975).
101. Meyer, G.-J., and Rössler, K., *Radiochem. Radioanal. Lett.* **25,** 377 (1976).
102. Meyer, G.-J., Rössler, K., and Stöcklin, G., *J. Labelled Comp Radiopharm.* **12,** 449 (1976).
103. Meyer, G.-J., and Lambrecht, R. M., *Int. J. Appl. Radiat.* Isot. **31,** 351 (1980).
104. Meyer, G.-J., Rössler, K., and Stöcklin, G., *Radiochim. Acta* **24,** 81 (1977).
105. Meyer, G.-J., Rössler, K., and Stöcklin, G., *J. Am. Chem. Soc.* **101,** 3121 (1979).
106. Mitchell, J. S., Brown, I., and Carpenter, R. N., *Experientia* **39,** 337 (1983).
107. Mitchell, J. S., Brown, I., and Carpenter, R. N., *Experientia* **41,** 925 (1985).
108. Mitchell, J. S., Brown, I., and Carpenter, R. N., "Human Alkaline Phosphatases" (T. Stigbrand and W. H. Fishman, Eds.). Liss, New York, 1985.
109. Mitchell, J. S., Brown, I., and Carpenter, R. N., "Strahlenschutz in Forschung [25 Jahre medizinischer Strahlenschutz]" (H. A. Ladner, C. Reiners, W. Borner, and J. Schutz, Eds.). Thieme, Stuttgart, 1985.
110. Nefedov, V. D., Norseev, Yu. V., Toropova, V. A., and Khalkin, V. A., *Russ. Chem. Rev.* (*Engl. Transl.*) **37,** 87 (1968).
111. Nefedov, V. D., Toropova, M. A., Khalkin, V. A., Norseev, Yu. V., and Kuzin, V. I., *Soviet Radiochem.* (*Engl. Transl.*) **12,** 176 (1970).
112. Nefedov, V. D., Norseev, Yu. V., Savlevich, Kh., Sinotova, E. N., Toropova, M. A., and Khalkin, V. A., *Dokl. Chem.* (*Engl. Transl.*) **142-147,** 507 (1962).
113. Neirinckx, R. D., Myburg, J. A., and Smit J. A., *Radiopharm. Labelled Compounds Proc. Symp. 1973*, Vol. 2, pp. 171–181 (1973).
114. Norseev, Yu. V., and Khalkin, V. A., *Chem. Zvesti* **21,** 602 (1967).
115. Norseev, Yu. V., and Nefedov, V. D., *Issled. Khim. Tekhnol. Primen. Radioakt. Veshchestv* **3** (1977).

116. Neumann, H. M., *J. Inorg. Nucl. Chem.* **4,** 349 (1957).
117. Perlman, I., Ghiorso, A., and Seaborg, G. T., *Phys. Rev.* **74,** 1730 (1948).
118. Perlman, I., Ghiorso, A., and Seaborg, G. T., *Phys. Rev.* **75,** 1096 (1949).
119. Persigehl, M. and Rössler, K., Deutscher Röntgenkongress, Berlin, AED-CONF-1975-904-029 (1975).
120. Ramler, W. J., Wing, D. J., Hendersen, D. J., and Huizenga, J. R., *Phys. Rev.* **114,** 154 (1959).
121. Ritter, M. A., Cleaver, J. W., and Tobias, C. A., *Nature (London)* **266,** 653 (1977).
122. Rössler, K., Tornau, W., and Stöcklin, G., *J. Radioanal. Chem.* **21,** 199 (1974).
123. Rössler, K., Meyer, G.-J., and Stöcklin, G., *J. Labelled Compd. Radiopharm.* **13,** 271 (1977).
124. Rössler, K., Meyer, G.-J., Feidendegen, L. E., and Stöcklin, G., "Nuklearmedizin" (K. Oeff and H. A. E. Schmidt, eds.). Medico informationdienste, Berlin, 1978.
125. Samson, G., Ph.D. Thesis, Frei Universiteit, Amsterdam, The Netherlands 1971.
126. Samson, G., and Aten, A. H. W., Jr., *Radiochim. Acta* **9,** 53 (1968).
127. Samson, G., and Aten, A. H. W., Jr., *Radiochim. Acta* **12,** 55 (1969).
128. Samson, G., and Aten, A. H. W., Jr., *Radiochim. Acta* **13,** 220 (1970).
129. Schultz, F., and Belleman, H., Kernforschungzentrum Karlsruhe, KFK 685 (1967).
130. Schutte, L., and Havinga, E., *Recl. Trav. Chim. Pays-Bas* **85,** 385 (1967).
131. Shaffer, R., *Trans. Am. Ophthamol.* Soc. 607 (1952).
132. Shellabarger, C. J., and Godwin, J. T., *J. Clin. Endocrinol. Metab.* **14,** 1149 (1954).
133. Shellabarger, C. J., Durbin, P. W., Parrott, M. W., and Hamilton, J. G., *Proc. Soc. Exp. Biol. Med.* **87,** 626 (1954).
134. Shiue, C.-Y., Meyer, G.-J., Ruth, T. J., and Wolf, A. P., *J. Labelled Compd. Radiopharm.* **18,** 1039 (1981).
135. Smit, J. A., Myburg, J. A., and Neirinckx, R. D., *Clin. Expol. Immunol.* **14,** 107 (1973).
136. Vasáros, L., Norseev, Yu. V., and Khalkin, V. A., Joint Institute for Nuclear Research, Dubra, U.S.S.R., JINR-12-12188 (1979).
137. Vasáros, L., Norseev, Yu. V., and Khalkin, V. A., Joint Institute for Nuclear Research, Dubra, U.S.S.R., JINR-P6-80-158 (1980).
138. Vasáros, L., Norseev, Yu. V., and Khalkin, V. A., Joint Institute for Nuclear Research, Dubra, U.S.S.R., JINR-P12-81-511 (1981).
139. Vasáros, L. Norseev, Yu. V., and Khalkin, V. A., *Dokl. Phys. Chem. (Engl. Transl.)* **262/267,** 161 (1982).
140. Vasáros, L., Norseev, Yu. V., and Khalkin, V. A., *Dokl. Phys. Chem. Engl. Transl.* **262/267,** 297 (1982).
141. Vasáros, L., Berei, K., Norseev, Yu. V., and Khalkin, V. A., *Magy. Kem. Foly.* **80,** 487 (1974).
142. Vasáros, L., Berei, K., Norseev, Yu. V., and Khalkin, V. A., *Radiochem. Radioanal. Lett.* **27,** 329 (1976).
143. Vasáros, L., Norseev, Yu. V., Nhan, D. D., and Khalkin, V. A., *Radiochem. Radional. Lett.* **47,** 313 (1981).
144. Vasáros, L., Norseev, Yu. V., Nhan, D. D., and Khalkin, V. A., *Radiochem. Radioanal. Lett.* **47,** 403 (1981).
145. Vasáros, L., Norseev, Yu. V., Berei, K., and Khalkin, V. A., *Radiochim. Acta* **31,** 75 (1982).
146. Vasáros, L., Norseev, Yu. V., Nhan, D. D., and Khalkin, V. A., *Radiochem. Radioanal. Lett.* **50,** 275 (1982).
147. Vasáros, L., Norseev, Yu. V., Forminykh, V. I., and Khalkin, V. A., *Soviet Radiochem. (Engl. Transl.)* **24,** 84 (1982).

148. Vasáros, L., Norseev, Yu. V., Nhan, D. D., and Khalkin, V. A., *Radiochem, Radioanal. Lett.* **54**, 239 (1982).
149. Vasáros, L., Norseev, Yu. V., Nhan, D. D., and Khalkin, V. A., *Radiochem. Radioanal. Lett.* **59**, 347 (1983).
150. Vasáros, L., Norseev, Yu. V., Meyer, G.-J., Berei, K., and Khalkin, V. A., *Radiochim. Acta* **26**, 171 (1979).
151. Vasáros, L., Norseev, Yu. V., Nhan, D. D., Khalkin, V. A., and Huan, N. Q., *J. Radioanal. Nucl. Chem. Lett.* **87**, 31 (1984).
152. Vaughan, A. T. M. Ph.D. Thesis, University of Birmingham, U.K., 1977.
153. Vaughan, A. T. M., *Int. J. Appl. Radiat. Isot.* **30**, 576 (1979).
154. Vaughan, A. T. M., *Int. J. Nucl. Med. Biol.* **7**, 80 (1980).
155. Vaughan, A. T. M., and Fremlin, J. H., *Int. J. Appl. Radiat. Isot.* **28**, 595 (1977).
156. Vaughan, A. T. M., and Fremlin, J. H., *Int. J. Nucl. Med. Biol.* **5**, 229 (1978).
157. Vaughan, A. T. M., Bateman, W., and Cowan, J., *J. Radioanal. Chem.* **64**, 33 (1981).
158. Vaughan, A. T. M., Bateman, W. J., and Fisher, D. R., *Int. J. Radiat. Oncol. Biol. Phys.* **8**, 1943 (1982).
159. Vaughan, A. T. M., Bateman, W. J., Brown, G., and Cowan, J., *Int. J. Nucl. Med. Biol.* **9**, 167 (1982).
160. Visser, G. W. M., Ph.D. Thesis, Frei Universiteit, Amsterdam, The Netherlands, 1982.
161. Visser, G. W. M., and Kaspersen, F. M., *Int. J. Nucl. Med. Biol.* **7**, 79 (1980).
162. Visser, G. W. M., and Diemer, E. L., *Int. J. Appl. Radiat. Isot.* **33**, 389 (1982).
163. Visser, G. W. M., and Diemer, E. L., *Radiochem. Radioanal. Lett.* **51**, 135 (1982).
164. Visser, G. W. M., Diemer, E. L., and Kaspersen, F. M., *Int. J. Appl. Radiat. Isot.* **30**, 749 (1979).
165. Visser, G. W. M., and Diemer, E. L., *Radiochim. Acta* **33**, 145 (1983).
166. Visser, G. W. M., Diemer, E. L., and Kaspersen, F. M., *J. Labelled Compd. Radiopharm.* **17**, 657 (1980).
167. Visser, G. W. M., Diemer, E. L., and Kaspersen, F. M., *Recl. Trav. Chim. Pays-Bas* **99**, 93 (1980).
168. Visser, G. W. M., Diemer, E. L., and Kaspersen, F. M., *Int. J. Appl. Radiat. Isot.* **31**, 275 (1980).
169. Visser, G. W. M., Diemer, E. L., and Kaspersen, F. M., *Int. J. Appl. Radiat. Isot.* **32**, 905 (1981).
170. Visser, G. W. M., Diemer, E. L., and Kaspersen, F. M., *J. Labelled Compd. Radiopharm.* **18**, 799 (1981).
171. Visser, G. W. M., Diemer, E. L., Vos, C. M., and Kaspersen, F. M., *Int. J. Appl. Radiat. Isot.* **32**, 913 (1981).
172. Visser, J., Brinkman, G. A., and Bakker, C. N. N., *Int. J. Appl. Radiat. Isot.* **30**, 745 (1979).
173. Walen, R. J., *J. Phys. Radium.* **10**, 95 (1949).
174. Wang Fu-Chiung, Kang Meng-Hua, and Khalkin, V. A., *Radiokhimiya* **4**, 94 (1962).
175. Whaling, W., "Handbuch der Physik" (S. Flügge, Ed.), Vol. 34, p. 211. Springer-Verlag, Berlin and New York, 1958.
176. Wolff, J., *Physiol. Rev.* **44**, 45 (1964).
177. Wolfgang, R., Baker, E. W., Caretto, A. A., Cumming, T. B., Friedlander, G., and Hudis, T., *Phys. Rev.* **103**, 394 (1956).
178. Zalutsky, M. R., Friedman, A. M., Buckingham, F. C., Wung, W., Stuart, F. P., and Simonian, S. J., *J. Lab. Comp. Radiopharm.* **13**, 181 (1977).
179. Milius, R. A., McLaughlin, W. H., Lambrecht, R. M., Wolf, A. P., Carroll, J. J., Adelstein, S. J. and Bloomer, W. D. *Appl. Radiat. Isotopes, Part A*, **37**, 799 (1986).
180. Eaborn, C., Najam, A. A. and Walton, D. R. M. *J. Chem. Soc. (Perkin)* **I**, 2481 (1972).

POLYSULFIDE COMPLEXES OF METALS

A. MÜLLER and E. DIEMANN

Fakultät für Chemie der Universität, D-4800 Bielefeld,
Federal Republic of Germany

I. Introduction

Many complexes with stable homonuclear diatomic ligands like O_2 and N_2 are well known and have been extensively studied. Only in recent years has the chemistry of metal complexes containing one or more coordinated homonuclear *sulfur ligands*, S_n^{2-} (with $n = 2$ or $n > 2$), been developed systematically. Complexes with S_n^{2-} units can be obtained for many metals under a variety of conditions, and coordinated S_n^{2-} ligands exhibit an especially rich structural chemistry.

Sulfide minerals were formed mostly hydrothermally from postmagnetic fluids; it appears remarkable that the formation of many ore deposits cannot be conclusively explained because of the very low solubility of the corresponding sulfide. Though it was postulated earlier (28, 29, 95) that polysulfido complexes might have been responsible for "the transport of metals and sulfur together" (80), it was later proved that many discrete polysulfido complexes and clusters of metals exist in polysulfide solutions (see, e.g., 136, 138).

The formation of soluble polysulfido clusters or complexes is also responsible for the fact that some classical analytical separation procedures based on polysulfide solutions (distinguishing between the classical thioanion-forming elements like As, Sb, Sn, and others like Cu) sometimes fail (51).

It is now evident that S_n^{2-} ions are fascinating and versatile (polydentate) ligands from the structural point of view, and that metal aggregates can be nicely "glued" by these ligands according to their high (and variable) number of coordination sites, which keeps the charge of the cluster low (134). In $[Cu_6(S_4)_3(S_5)]^{2-}$ only four ligands are capable of stabilizing a cluster with six metal atoms (134). It is remarkable that all possible S_n^{2-} ions ($n = 2$–9) occur in complexes, although S_9^{2-} has not been reported until now as an isolated ion.

It should also be noted that polysulfido complexes are not only interesting because of their structures and reactivity, but also because of their possible applications. They can, for example, be used to prepare sulfur rings of predetermined size and they are also suspected to play a role in catalysis (particularly in hydrodesulfurization).

II. Polysulfide Ions and Solutions

Free polysulfide ions consist of sulfur chains. The atoms of an S_3^{2-} chain are necessarily coplanar. Longer chains, however, exhibit different possibilities of isomerism. For example, by addition of one sulfur atom d- and l-S_4^{2-} can be obtained (Fig. 1). To describe the structures, in addition to bond lengths and interbond angles, the dihedral angle must be specified. For polysulfides, this angle is found to be between 60 and 110° in various examples. In order to illustrate the isomerism, the interbond and dihedral angles in Fig. 1 have been idealized to 90°.

The S_5^{2-} chain derives from the S_4^{2-} one by addition of one more sulfur atom, so that the dihedral angle of 90° is maintained. In doing so, we obtain structures c and d (Fig. 1), the cis and (d and l) trans forms of the S_5^{2-} chain. Similarly, by further addition of one sulfur atom, it can be shown that for an S_6^{2-} chain three (enantiomorphic) isomers will result (cis,cis; trans,trans; and cis,trans, each d and l, respectively). The cis-S_5^{2-} and cis,cis-S_6^{2-} ions may be regarded as parts of the S_8 ring

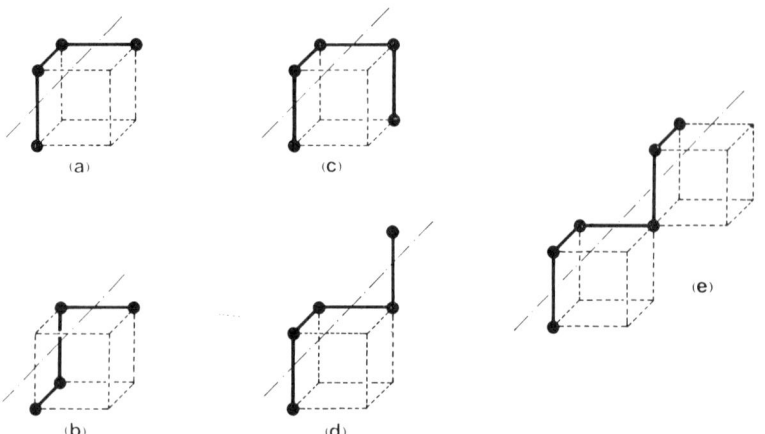

FIG. 1. Idealized stereochemistry (dihedral angle equal to 90°) for polysulfide ions (*191*). (a), (b), d- and l-S_4^{2-}; (c), (d), cis- and $trans$-S_5^{2-}; (e), $trans,trans$-S_6^{2-}

structure, while $trans$-S_5^{2-} and $trans,trans$-S_6^{2-} correspond to portions of an infinite helical chain (like that, e.g., in fibrous sulfur). All-trans conformations of the polysulfide chains have been found in, for example, Cs_2S_5 (*10*), Cs_2S_6 (*8*), and $(PPh_4)_2S_7$ (*70*); this is obviously the normal arrangement. However, the cis conformation has been detected in α-Na_2S_5 (*9*), the cis,trans conformation in $[(CH_3)_4N](NH_4)S_6$ (*112*), and the cis conformation for the five central sulfur atoms of S_7^{2-} in $(PPh_4)(NH_4)S_7 \cdot CH_3CN$ (*126*).

Solutions containing several S_n^{2-} species (but mainly S_4^{2-} and S_5^{2-}) are obtained upon digesting sulfur with aqueous sulfide. Solutions in methanol and ethanol may be prepared correspondingly. Salts of S_n^{2-} ($n = 2$–8) can be obtained not only from such solutions but also from liquid NH_3 and in dry reactions.

The average S—S bond length is shorter in S_n^{2-} ($n > 2$) than in S_2^{2-}, and the S—S terminal bond lengths decrease from S_3^{2-} (2.15 Å) to S_7^{2-} (1.992 Å) (*70*). These facts indicate that the negative charge (filling a π*-antibonding molecular orbital) is delocalized over the whole chain, but the delocalization along the chain is less in higher polysulfides. This consideration is significant in comparing the S—S bond length of the free ions with those of the coordinated ones.

III. Survey of Compound Types

A. COMPLEXES WITH DIFFERENT TYPES OF COORDINATED S_2^{2-} LIGANDS

Complexes with S_2^{2-} ligands show a remarkable variety of structures, which result from the extension of the fundamental structural type Ia (side-on coordination) as well as IIa and IIb [cis and trans "end-on" (doubly) bridging coordination],[1] by using the remaining lone pairs of electrons (on sulfur) to coordinate additional metal atoms. Such coordination of additional metal atoms occurs for all three fundamental structural types (Ia, IIa, IIb). Representative examples (including structural type definitions) are summarized in Table I. End-on coordination of an S_2^{2-} ligand to a single metal atom to give a terminal MS_2 unit is unusual and has been claimed only for matrix-isolated MgS_2, ZnS_2, and CdS_2 (*94*).

Another fundamental type of structural unit found in many cluster compounds is III, in which both sulfur atoms of the ligand are bonded to

---
[1] These types are also known for dioxygen complexes (*88, 186*).

TABLE I

Typical Geometries of S_2^{2-} Complexes[a]

Structural type	Example	Structural type	Example
Ia M⟨S-S⟩M (triangle)	[Ir(dppe)$_2$(S$_2$)]$^+$ (1)	IIc S-S with M, M, M	Cp$_4$Co$_4$(S$_2$)$_2$S$_2$
Ib M-S-S-M (with M above)	[Mo$_4$(NO)$_4$(S$_2$)$_5$S$_3$]$^{4-}$ (4)	IId M, M-S-S-M, M	(SCo$_3$(CO)$_7$)$_2$(S$_2$) (9)
Ic M-S-S-M	Mn$_4$(S$_2$)$_2$(CO)$_{15}$ (6)	IIIa S-S with M, M, M	[Mo$_2$(S$_2$)$_6$]$^{2-}$ (10)
Id M-S-S-M (bridged)	Mn$_4$(S$_2$)$_2$(CO)$_{15}$ (6)	IIIb M-S-S-M diamond	[(triphos)$_2$Ni$_2$(S$_2$)]$^+$
IIa M-S-S-M	[(NH$_3$)$_5$Ru(S$_2$)Ru(NH$_3$)$_5$]$^{4+}$	IV M-S-S-M	MeCp$_2$Cr$_2$(S$_2$)$_2$S (12)
IIb M-S-S-M	[Mo$_4$(NO)$_4$(S$_2$)$_6$O]$^{2-}$ (8)		

[a] For references see text and Table II.

each of two metal atoms. The S—S bond is oriented approximately perpendicular to the metal–metal vector. A very remarkable, but uncommon, type of bridging ligand has been discovered in which two metal atoms are attached to the same sulfur atom of the ligand (structure type IV) (*16*).

1. Type I Complexes

Complexes with type I structures are listed in Table II along with their S—S distances [d(S—S)] and S—S vibrational frequencies [v(S—S)]. The most common mode of coordination is structure type Ia, in which the ligand occupies two coordination sites of the metal atom. One example of this type of binding is the complex [Ir(dppe)$_2$(S$_2$)]Cl (**1**) (*7, 55*). The Ir atom is in a distorted octahedral environment, as would be expected for a six-coordinate metal with a d^6 electron configuration. Higher coordination numbers are found for complexes of the early transition elements. In [MoO(S$_2$)$_2$(C$_2$O$_3$S)]$^{2-}$ (**2**) the Mo atom adopts pentagonal-bipyramidal coordination geometry with the S atoms of S$_2^{2-}$ and of the thiooxalate ligand lying in the pentagonal plane (*100, 101*).

Several binuclear Mo(V) complexes contain type Ia ligands. The central {MoO(η^2-S)$_2$MoO}$^{2+}$ structural unit in (NMe$_4$)$_2$[Mo$_2$O$_2$S$_2$(S$_2$)$_2$] (**3**) has the Mo=O groups in syn stereochemistry (*26, 161*). Related examples are [Mo$_2$OS$_7$]$^{2-}$ (where one oxygen of **3** is substituted by sulfur) (*139*) and [W$_2$S$_{10}$]$^{2-}$ (with one S$_4^{2-}$ ligand at one tungsten atom) (*139*).

Ligands of structure type Ib also exist in the interesting compound (NH$_4$)$_4$[Mo$_4$(NO)$_4$(S$_2$)$_5$S$_3$]·2H$_2$O (**4**), which contains sulfur atoms in five different bonding situations (*118*). Four of the five S$_2^{2-}$ ligands

TABLE II

Representative Examples of Complexes with Types I–III S_2^{2-} Ligands

Formula	Type	$d(S-S)^a$ (Å)	$\nu(S-S)^a$ (cm^{-1})	Color	Reference
[Ir(dppe)$_2$(S$_2$)]$^+$	Ia	2.07	528	Orange	7, 55
[Rh(dmpe)$_2$(S$_2$)]$^+$	Ia	—	525	Orange	55
RhL1(S$_2$)Cl^b	Ia	—	546	Orange	151, 152
Os(CO)$_2$(PPh$_3$)$_2$(S$_2$)	Ia	—	—	Orange	23, 47
[Rh(vdiars)$_2$(S$_2$)]$^{+c}$	Ia	—	554	Red-brown	89
[Rh(L2)$_2$(S$_2$)]$^{+d}$	Ia	—	—	Brown	46
[MoO(S$_2$)$_2$(mtox)]$^{2-e}$	Ia	2.01	530	Dark red	100, 101
MoO(S$_2$)(dtc)$_2$^{f}	Ia	2.02	558	Blue	36, 37
Cp$_2$Mo(S$_2$)	Ia	—	536	Red	77
Cp$_2$Nb(S$_2$)Cl	Ia	—	540	Red	184
Cp$_2$Nb(S$_2$)Me	Ia	2.01	540	Orange	1, 102
[Mo$_2$O$_2$S$_2$(S$_2$)$_2$]$^{2-}$	Ia	2.08	510	Red-orange	26, 27, 161
[Mo$_2$S$_4$(S$_2$)(S$_4$)]$^{2-}$	Ia	2.07	—	Red-violet	25
[Mo$_4$(NO)$_4$(S$_2$)$_5$S$_3$]$^{4-}$	Ia	2.04	536	Red	118
[Mo$_4$(NO)$_4$(S$_2$)$_5$S$_3$]$^{4-}$	Ib	2.05	550	Red	118
Cp$_2$Fe$_2$(S$_2$)$_2$CO	Ib	1.99	—	Green	52
Mn$_4$(S$_2$)$_2$(CO)$_{15}$	Ic	2.07	—	Red	84, 85
Mn$_4$(S$_2$)$_2$(CO)$_{15}$	Id	2.09	—	Red	84, 85
[(CN)$_5$Co(S$_2$)Co(CN)$_5$]$^{6-}$	IIa	—	490	Red-brown	174
[(NH$_3$)$_5$Ru(S$_2$)Ru(NH$_3$)$_5$]$^{4+}$	IIa	2.01	514	Green	15, 44
Cp(CO)$_2$Mn(S$_2$)MnCp(CO)$_2$	IIa	2.01	—	Dark green	62
{(Re$_6$S$_8$)S$_{4/2}$(S$_2$)$_{2/2}$$^{4-}$}$_n$	IIa	2.09	—	Dark red	14, 22, 171
Cp$_2$Fe$_2$(S$_2$)(SEt)$_2$	IIb	2.02	507	Dark green	82, 83, 183
[Mo$_4$(NO)$_4$(S$_2$)$_6$O]$^{2-}$	IIb	2.08	480	Violet	115
Cp′$_2$Mo$_2$(S$_2$)$_5$^{g}	IIb^h	2.04	—	Black	158
Cp$_4$Co$_4$(S$_2$)$_2$S$_2$	IIc	2.01	—	Black	185
Cp$_4$Fe$_4$(S$_2$)$_2$S$_2$	IIc	2.04	503	Black	82, 187
(SCo$_3$(CO)$_7$)$_2$(S$_2$)	IId	2.04	—	Black	93, 180
{CpMn(NO)(S$_2$)}$_n$	IId	—	491(?)	Red-brown	154
[(Cp$_4$Fe$_4$(S$_2$)$_2$S$_2$)$_2$Ag]$^+$	IId	2.05	478	Black	82, 187
[Mo$_2$(S$_2$)$_6$]$^{2-}$	IIIa	2.04	550	Green-black	127, 128
{Mo$_2$(S$_2$)$_2$Cl$_4$Cl$_{4/2}$}$_n$	IIIa	1.98	561	Dark brown	92
Nb$_2$(S$_2$)$_2$Cl$_4$	IIIa	2.03	588	Brown	155, 170
[Mo$_3$S(S$_2$)$_6$]$^{2-}$	IIIa	2.02	545	Red	129, 140
Mo$_3$S(S$_2$)$_3$Cl$_4$	IIIa	2.03	562	Red	92
Fe$_2$(S$_2$)(CO)$_6$	IIIa	2.01	555	Red	63, 71, 190
Ta$_2$(S$_2$)$_2$(PS$_4$)$_2$	IIIa	2.05	—	Grey	49
Mo$_2$(n-BuCp)$_2$(S$_2$)Cl$_4$	IIIa	2.02	—	Black	17, 103
Mo$_2$(S$_2$)(S$_2$C$_2$Ph$_2$)$_4$	IIIa	2.04	518	Green-black	193
Nb$_2$S(S$_2$)Br$_4$(tht)$_4$^{i}	IIIa	2.01	—	Green	41
[Mo$_2$(S$_2$)(SO$_2$)(CN)$_8$]$^{4-}$	IIIa	2.00	520	Violet	157
[(triphos)$_2$Ni$_2$(S$_2$)]$^+$	IIIb	2.21	—	Brown	99

^{a} Mean value.
^{b} L1 = PPh(CH$_2$CH$_2$CH$_2$PPh$_2$)$_2$.
^{c} vdiars = Ph$_2$AsCHCHAsPh$_2$.
^{d} L2 = R$_2$PNHPR$_2$.
^{e} mtox = O$_2$CCOS^{2-}.
^{f} dtc = S$_2$CNPr$_2$.
^{g} Cp′ = Me$_5$Cp.
^{h} Mo—S—S—Mo angle ≈ 59.7°.
^{i} tht = tetrahydrothiophene.

bridge pairs of Mo atoms unsymmetrically (structure type Ib). One S_2^{2-} ligand is "side-on" coordinated to a single Mo atom (structure type Ia).

It is also interesting to note that in $[Cl_2FeS_2MoO(S_2)]^{2-}$ (**5a**) (*141*) and $[S_2WS_2(WS)S_2(WS)(S_2)]^{2-}$ (**5b**) (*194*) (with type Ia ligands) the units $\{S_2MoO(S_2)\}^{2-}$ and $\{S_2WS(S_2)\}^{2-}$ may be regarded as perthio derivatives of the thiometalates $MoOS_3^{2-}$ and WS_4^{2-}, respectively.

Structure types Ic and Id (Table I), in which the ligand is coordinated to three and four metal atoms, respectively, both occur in $Mn_4(S_2)_2(CO)_{15}$ (**6**) (*84, 85*).

2. Complexes with Type II Structure

Complexes with type II structures are compiled in Table II along with their S—S distances and S—S vibrational frequencies. There are several examples of complexes of structure type IIa (planar trans end-on bridging coordination). Vibrational spectroscopy (*174*) indicates that this structure occurs in $[(CN)_5Co(S_2)Co(CN)_5]^{6-}$ (**7**). Trans arrangement of the metal atoms has been proved by X-ray structure determination for $[(NH_3)_5Ru(S_2)Ru(NH_3)_5]Cl_4 \cdot 2H_2O$ (*15, 44*).

Structure type IIb (cis end-on bridging coordination) is found in $Cp_2Fe_2(S_2)(SEt)_2$ (*83, 183*), where it is dictated by the two bridging SEt groups. The Fe(S$_2$)Fe group is planar (as in type IIa structures). The same kind of coordination occurs in $[Mo_4(NO)_4(S_2)_6O]^{2-}$ (**8**) (*115*). This complex represents a tetragonal bisphenoid of Mo atoms with an interstitial oxygen atom. The two nonadjacent edges of the metal cage are bridged by type IIb S_2^{2-} ligands. The remaining four edges are bridged by ligands of structure type IIIa (*vide infra*).

The complex $Cp_4Co_4(S_2)_2S_2$ has a cage structure with two S_2^{2-} ligands, each of which bridges three Co atoms (type IIc) (*185*). A few complexes have been structurally characterized in which an S_2^{2-} ligand bridges four metal atoms (structure type IId, Tables I and II). In the cluster compound $(SCo_3(CO)_7)_2(S_2)$ (**9**), the ligand is bound to one edge of each of two Co$_3$ triangles (*93, 180*). The resulting structure consists of two Co$_3$S$_2$ planes which have a common S_2^{2-} edge.

3. Complexes with Type III Structures

Type III complexes are included in Table II. The binuclear compound $(NH_4)_2[Mo_2(S_2)_6]\cdot 2H_2O$ (**10**) incorporates only S_2^{2-} ligands (*127, 128*). Two of the S_2^{2-} ligands bridge the two Mo atoms (type IIIa) and four of the S_2^{2-} ligands are bonded to a single Mo atom (type Ia geometry). The coordination geometry about each Mo atom is distorted dodecahedral. An interesting feature of the type IIIa bridging S_2^{2-} ligands is their asymmetric bonding to the Mo atoms. A second structural isomer of $[Mo_2S_{12}]^{2-}$, with two terminal and two bridging sulfides and one S_4^{2-} ligand at each Mo atom, and the corresponding tungsten species $[W_2S_4(S_4)_2]^{2-}$ (**30b**), have also been reported (*40, 139*).

Another cluster anion with type IIIa ligands which contains only molybdenum and sulfur occurs in $(NH_4)_2[Mo_3S(S_2)_6]\cdot nH_2O$ (**11**) (*111, 129*) ($n = 0$–2, with variable nonstoichiometric amounts of water, which are disordered in the crystal lattice) (*35*). This type of ligand has also been found in a compound containing two isostructural anions in the crystal lattice, i.e., the mixed valence $[(Mo^{III/IV})_2(SO_2)(S_2)(CN)_8]^{5-}$ (paramagnetic) and $[Mo_2^{IV}(SO_2)(S_2)(CN)_8]^{4-}$ (diamagnetic) (117).

Structure type IIIb with an η^2-S_2^{2-} ligand that is coplanar with the two metal atoms has been detected in [(triphos)Ni(μ-S$_2$)-Ni(triphos)]ClO$_4$[triphos = bis(2-diphenylphosphinoethyl)phenylphosphine] (99).

In $Cp'_2Cr_2S(S_2)_2$ (12) two Cr atoms are bridged by a type IIIa S_2^{2-} ligand, by a sulfido group, and by a type IV ligand (16). This compound is the first known example of the latter type. Both the type IIIa and type IV S—S bonds are surprisingly long (2.15 and 2.10 Å, respectively).

B. Complexes with Ring Systems Generated by S_n^{2-} Ions ($n > 2$)

Different types of coordination of S_n^{2-} ligands are shown in Fig. 2.

1. Mononuclear Complexes with $M(\mu_2$-$S)_n$ Ring Systems (Bidentate Ligands)

Only a few compounds containing the S_3^{2-} ligand are known. The compound (MeCp)$_2$Ti(S$_3$) (13) contains a nonplanar four-membered TiS$_3$ ring with a dihedral angle of 49° between TiS(1)S(2) and

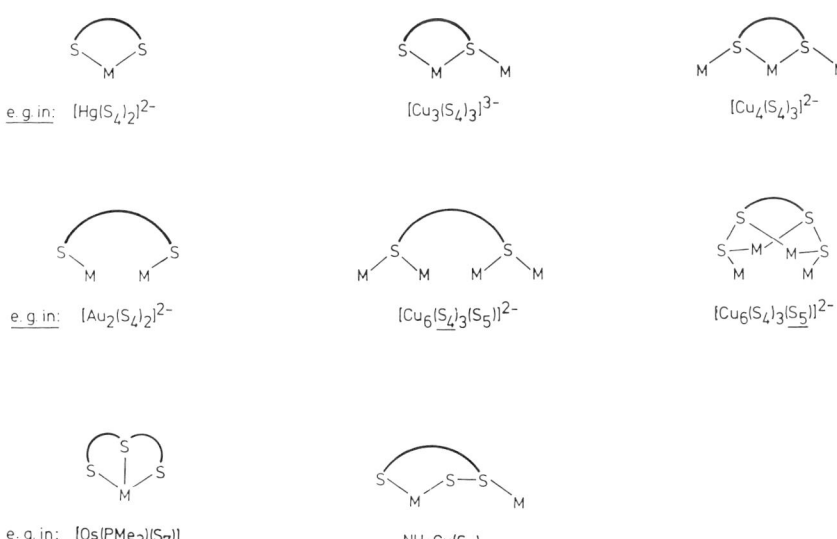

FIG. 2. Important structure types observed for S_n^{2-} complexes (mainly with $n = 4$).

S(1)S(2)S(3) (2). On the other hand quite a number of complexes with S_4^{2-} ligands have been prepared and structurally characterized. In compounds where the ion acts as a bidentate ligand, an "envelope" as well as a "half-chair" conformation is possible. In $Cp_2M(S_4)$ (M = Mo, W) (complexes with tetrahedral coordination of the metal ions) (34, 73) and in $[M(S_4)_2]^{2-}$ [M = Ni (**14**), Pd] the half-chair conformation is found (123) whereas several complexes with a square-pyramidal coordination of Mo and W show the envelope conformation (25, 40, 139). The S_4^{2-} ligand has also been found in $[Sn(S_4)_3]^{2-}$, $[Sn(S_4)_2(S_6)]^{2-}$ (143), and $[M(S_4)_2]^{2-}$ [M = Zn (31), Hg (**17b**) (142)].

Quite a few complexes with the bidentate pentasulfido ligand are also known. The first reported was the homoleptic and optically active complex $[Pt(S_5)_3]^{2-}$ (**15**) (53, 64, 65, 68, 69, 176). Brick-red $(NH_4)_2[Pt(S_5)_3] \cdot 2H_2O$ is formed from the reaction of $K_2[PtCl_6]$ with aqueous $(NH_4)_2S_n$ solution. Addition of concentrated HCl results in the separation of maroon $(NH_4)_2[PtS_{17}] \cdot 2H_2O$ (54). The $[Pt(S_5)_3]^{2-}$ ion crystallizes from the solution as a racemate, which can be resolved by forming diastereoisomers. Upon crystallization, $[PtS_{17}]^{2-}$ undergoes a second-order asymmetric transformation, so that the solid contains an excess of the (−) enantiomer (54).

13   **14**   **15**   **16**

17a   **17b**   **18**

The compound $Cp_2Ti(S_5)$ (76, 167), which has been prepared by various different routes (*vide infra*), is probably the most thoroughly studied polysulfido compound. Some of the more important reactions of this complex will be discussed below. The corresponding compounds of zirconium, hafnium, and vanadium have been obtained by the reaction of Cp_2MCl_2 with $Li_2S_2/3S$ or $(NH_4)_2S_5$, respectively (78, 97). The spectroscopic properties of the black paramagnetic complex $Cp_2V(S_5)$ are particularly interesting and helpful for a general description of the electronic structure of Cp_2ML_2 compounds (156).

The MS_5 moiety has the chair conformation in all known mononuclear complexes. Examples are $[Cr(NH_3)_2(S_5)_2]^{2-}$ (16) (143) and $[(S_5)Mn(S_6)]^{2-}$ (31). It is worth noting that the "bite" of S_5^{2-} varies strongly [a very large one is found in $[(S_5)Fe(MoS_4)]^{2-}$ (30, 32)].

Some mononuclear complexes containing bidentate S_6^{2-} ligands can also be isolated. Examples include the homoleptic $[M(S_6)_2]^{2-}$ [M = Zn, Cd, Hg (17a)] (144, 145) and two containing the S_9^{2-} ligand, namely $[AuS_9]^-$ (91) and $[AgS_9]^-$ (18) (137, 143), both with ring structure. The following homoleptic mononuclear complexes are known at present: $[M(S_4)]^{2-}$ [M = Zn (31), Ni (123), Pd (123), Hg (142)], $[Pt(S_5)_2]^{2-}$ (168, 192), $[M(S_5)_3]^{m-}$ [M = Pt, m = 2 (53, 64, 65, 69); M = Rh, m = 3 (79)], $[M(S_6)_2]^{2-}$ [M = Zn, Cd (142), Hg (144)], and $[M(S_9)]^-$ [M = Ag (137, 143), Au (91)].

Monocyclic MS_n units are known for $n = 3–7$ and 9. Among the mononuclear complexes there is also an example where a polysulfide acts as a tridentate ligand, namely the S_7^{2-} ion in $(Me_3P)_3M(S_7)$ (M = Ru, Os). The OsS_7 bicycle shows a similar exo-endo conformation as the S_8^{2+} ion in $S_8(AsF_6)_2$ (58).*

2. Binuclear Complexes with $M(\mu_2 - S)_n M$ Units (Doubly Bridging Ligands)

Only a few binuclear complexes are known in which S_n^{2-} ($n = 3–8$) acts as a doubly bridging ligand. The structure of $(MeCp)_4Ti_2S_6$ (19) consists of an eight-membered ring containing alternating Ti atoms and S_3 fragments. In contrast to cyclo-S_8, the ring adopts a "cradle" conformation with the Ti atoms positioned at the sites adjacent to the apical S atoms. In the complex, the S—S—S angles are expanded relative to those known for other cyclic sulfur ring species (6).

The $\{S_5Mo_2(S')\}$ moiety in $[Mo_2(NO)_2(S_2)_3(S_5)(OH)]^{3-}$ (20a), however, has practically the same geometry as that of cyclooctasulfur [if one ring member is assumed to be at the center of the bridging S_2^{2-} group ("S'")] (114). The highest known oxidation state (+5) of a metal atom in a polysulfido (S_n^{2-}, $n > 2$) was found in $[Nb_2(OMe)_2(S_2)_3(S_5)O]^{2-}$ (20b), in which the $\{Nb_2S'S_5\}$ fragment has the conformation of S_8^{2+} (143). The compound $Os_2\text{-}\mu\text{-}(S_5)\text{-}\mu\text{-}(S_3CNR_2)\text{-}(S_2CNR_2)_3$ (21) contains the S_5^{2-} ion as a "half-bridging" ligand (90).

---

* Polynuclear species with monocyclic MS_4 entities such as $W_2S_{12}^{2-}$ (30b) are not discussed separately.

The six-membered OsS$_5$ ring has the chair conformation. A seven-membered 1,4-dirheniacycloheptasulfur ring has been found in Cp$_2$Re$_2$(CO)$_2$(μ-S$_2$)(μ-S$_3$) (**22**) (*61*).

The remarkable 10-membered highly symmetrical [Au$_2$S$_8$]$^{2-}$ ring (**23**) consists of two Au atoms and two bridging S$_4^{2-}$ ligands and has approximately D$_2$ symmetry (*136*). The binuclear complexes

$[(S_6)M(S_8)M(S_6)]^{4-}$ (M = Cu (**24**), Ag) have two bidentate S_6^{2-} ligands (forming rings) and the doubly bridging S_8^{2-} ligand (*109, 133*). A novel bismuth polysulfido compound with the highest portion of sulfur known so far, $[Bi_2(S_6)(S_7)_4]^{4-}$ (**25a**), has also been prepared (*149*).

An interesting example of doubly bridging S_7^{2-} ligands has been found in $[Pd_2(S_7)_4]^{2-}$ (**25b**) (*148*) (with a NH_4^+ ion captured in the center of the cage).

3. Polynuclear Complexes with Condensed Ring Systems (Polydentate Ligands)

One would expect (according to the high number of coordination sites) that "soft metal aggregates" could be "glued" by S_n^{2-} ions.

A remarkable species with very different types of coordination is the hexanuclear complex $[Cu_6S_{17}]^{2-}$ (**26**) containing 10 μ_3-sulfur atoms and an aggregate of six Cu(I) ions which can be approximately described as two Cu_4 tetrahedra sharing one edge (*134, 135*). Whereas only the two terminal sulfur atoms of the three S_4^{2-} ligands in **26** are bonded to copper atoms, coordination of four sulfur atoms of the S_5^{2-} ligand is found. Each of the two equivalent S_4^{2-} ions acts as a bridging and chelating ligand for the two "Cu tetrahedra," with coordination to three different Cu atoms. The S4 and S7 atoms (correspondingly S4' and S7') are bonded not only to the same atom Cu1 (Cu1'), but also to Cu3 and Cu2' (Cu3' and Cu2), respectively. The other S_4^{2-} (S8—S9—S9'—S8') and the S_5^{2-} ligands are responsible for the connection of the "tetrahedra." Both ligands show coordination to four different Cu atoms.

The tetranuclear clusters $[Cu_4(S_n)_3]^{2-}$ ($n = 4$ (**27a**), 5) [including $[Cu_4(S_5)_2(S_4)]^{2-}$ and $[Cu_4(S_5)(S_4)_2]^{2-}$] have six μ_3-S atoms from three S_n^{2-} ligands as bridges for the six edges of the metal tetrahedra. The trinuclear species $[Cu_3(S_6)_3]^{3-}$ (**27b**) and $[Cu_3(S_4)_3]^{3-}$ have three μ_3-S atoms forming the central Cu_3S_3 ring with the chair conformation (*109, 135, 136*).

The basic structure of the species $[Cu_4(S_n)_3]^{2-}$ can formally be obtained from $[Cu_3(S_n)_3]^{3-}$ (a novel polycyclic inorganic species with different puckered copper sulfur heterocycles, a central Cu_3S_3 ring with the chair conformation, and three outer CuS_6 or CuS_4 rings) by coordination of each of the three "end-on bonded sulfur atoms" of the three polysulfide ligands to an additional fourth Cu(I) (*135, 136*). The structure of $[Cu_6S_{17}]^{2-}$ can formally be derived by connecting two $Cu_3(\mu\text{-}S)_3$ rings by four polysulfido ligands.

It turns out that the six-membered $Cu_3(\mu\text{-}S)_3$ rings are paradigmatic units. This type of ring system has been incorporated into current models of metallothioneins [low-molecular-weight proteins which are believed to play a key role in metal metabolism (cf. references in *136*)]. The structural chemistry of the Ag complexes seems to be different. Monocyclic $\{Ag(S_n)\}^-$ rings can be linked via bridging ligands as in $[(S_6)Ag(S_8)Ag(S_6)]^{4-}$ (*133*) or condensed as in $[Ag_2(S_6)_2]^{2-}$ (**28**) (*126*).

4. Ring Systems with Strong Metal–Metal Bonds

The cubane cluster $[Re_4S_4(S_3)_6]^{4-}$ (**29a**) with six bridging S_3^{2-} ligands has been reported recently (*124*). This is the only known species where metal–sulfur rings with strong metal-to-metal bonds occur. The six Re_2S_3 rings have planar Re_2S_2 moieties and, therefore, an envelope conformation. Complexes of the type $[Re_4S_4(S_3)_n(S_4)_{6-n}]^{4-}$ can also be obtained (*107*).

<u>29a</u>   <u>29b</u>

5. Mixed (Polysulfido) Ligand Complexes

Several complexes, particularly the labile ones discussed above, contain two different S_n^{2-} ligands (e.g., **27a**). Another nice example is $CpTi(S_2)(S_5)$ (**29b**) (*107*) with a pseudo-sandwich structure.

C. Solid-State Structures

Many solid-state structures with S_n^{2-} groups are known (e.g., $Na_2Re_3S_6$, NbS_2Cl_2, MoS_2Cl_3, or $Mo_3S_7Cl_4$; all having S_2^{2-} ligands). The reviews of Rabenau and co-workers (*48*) and of Bronger (*13*) cover the most important literature in this field. Regarding the ability of S_n^{2-} species to glue metal aggregates together we restrict ourselves here to two characteristic examples. The complex $(NH_4)_2PdS_{11} \cdot 2H_2O$ (*60*) contains Pd atoms linked via S_6^{2-} chains in a three-dimensional array. The crystal structure is a composite of different possible linkages, whereby sulfur absences account for the PdS_{11} composition. In $(NH_4)Cu(S_4)$, well known from laboratory courses, CuS_4 chelate rings are linked via "additional" Cu—S bonds to form one-dimensional polymeric anions (*18*).

IV. Syntheses

There are several routes for the synthesis of polysulfido complexes. Oxidative addition of elemental sulfur to a coordinatively unsaturated electron-rich metal is a convenient method for preparing S_n^{2-} complexes. Examples are the reactions of $CpRh(PPh_3)_2$ to yield $CpRh(PPh_3)(S_5)$ (*189*) or of ML_4 (M = Pd, Pt; L = PPh_3) to yield $L_2M(S_4)$ (*21*). Examples of the preparation of complexes by using sulfur as reagent are the preparation of $Cp_2Ti(S_5)$ from $Cp_2Ti(CH_3)_2$ or $Cp_2Ti(SH)_2$ (*76, 163*) and the following reactions (*77*), which proceed without a change in the oxidation state of Mo (see also section V).

$$Cp_2Mo(S_2) + \tfrac{1}{4}S_8 \longrightarrow Cp_2Mo(S_4)$$

$$Cp_2Mo(SH)_2 + \tfrac{3}{8}S_8 \longrightarrow Cp_2Mo(S_4) + H_2S$$

By reacting thioanions of molybdenum, tungsten, or rhenium (e.g., MoS_4^{2-}, $MoOS_3^{2-}$, WS_4^{2-}, and ReS_4^-) with sulfur, complexes containing MS_4 ring systems like $[SM(S_4)_2]^{n-}$ (**30a**), $[OMo(S_4)_2]^{2-}$ (**30a**) (M = Mo, Re), $[Mo_2O_2S_2(S_4)_2]^{2-}$, and $[W_2S_4(S_4)_2]^{2-}$ (**30b**) have been obtained (*25, 40, 107, 139*).

```
  X
  |
S-S\ | /S-S
 \ \ M / /         M=Mo,Re
 / /   \ \         X=O,S
S-S     S-S
    30a
```

```
    S   S   S
    |   |   |
S-S\ \ / \ / /S-S
 \  \ W   W  /  \
 /  / \ / \ \   /
S-S    S    S-S
       30b
```

Several compounds with network structures have been prepared, by high-temperature reaction of metals or metal halides with S_8 and/or S_2Cl_2, as for example in the reaction of a metal halide with sulfur and S_2Cl_2 (*92*) (see also Section V).

$$MoCl_3 + S_8/S_2Cl_2 \longrightarrow Mo_3S(S_2)_3Cl_4 + Mo_2(S_2)_2Cl_6$$

Reaction of metal complexes with S_n^{2-} is a convenient method for directly introducing the ligand by substitution of other ligands. For example, Na_2S_2 or polysulfide solutions can be used for this purpose (*77, 107, 108, 120, 146*).

$$Cp_2MoCl_2 + Na_2S_2 \longrightarrow Cp_2Mo(S_2) + 2NaCl$$
$$[Mo_3S_4(CN)_9]^{5-} + S_n^{2-} \longrightarrow [Mo_3S(S_2)_6]^{2-}$$
$$Mo_3S(S_2)_3Cl_4 + S_n^{2-} \longrightarrow [Mo_3S(S_2)_6]^{2-}$$
$$FeCl_3 + S_n^{2-} \longrightarrow [Fe_2S_2(S_5)_2]^{2-}$$
$$MoO_4^{2-} + S_n^{2-} \longrightarrow [Mo_3S(S_2)_6]^{2-} + [Mo_2(S_2)_6]^{2-}$$

In the latter reaction, S_n^{2-} acts as a reducing agent.

Nearly all homoleptic complexes have been prepared from reactions with polysulfide solutions. Solvents used in most cases were H_2O, CH_3OH, C_2H_5OH, CH_3CN, or dimethylformamide (DMF). It seems that the type of S_n^{2-} ligand occurring in the complex need not necessarily have a high abundance in the polysulfide reaction solution: Cp_2TiCl_2 reacts with Na_2S_n ($n = 2$–7) to yield $Cp_2Ti(S_5)$ in all cases (*75*), i.e., the most stable M_yS_n system is formed. This consideration applies also to the very simple reaction of metal ions with H_2S in the presence of oxygen, where the formation of only small amounts of a "matching" ligand suffices for the preparation of a particular complex, like $[Fe_2S_2(S_5)_2]^{2-}$ (*107*).

On the other hand, the compounds in the system $Cu(I)/S_n^{2-}$ are obviously kinetically labile. Here, *n* can be influenced simply by passing H_2S through the S_8-containing reaction mixture (keeping all other conditions constant). Chains with high *n* are formed when the portion of S^{2-} (from H_2S) is low. With decreasing S_8/S^{2-} ratio, the chain size decreases. By controlling this ratio a series of different copper clusters has been obtained (see Section III,B,3) (107).

The formation of polysulfido complexes with other reagents containing S_n bonds, such as Cl_2S_n and R_2S_n, is also possible (43, 74). The former offers a particularly interesting reaction (43).

$$\text{>M<}_{S}^{S} + Cl_2S_2 \longrightarrow S \underset{S-S}{\overset{>M<}{\diagdown\diagup}} S + 2Cl^-$$

It has been pointed out (33) that synthesis of $[Fe_2S_2(S_5)_2]^{2-}$ from $[Fe(SPh)_4]^{2-}$ and dibenzyl trisulfide would have implications regarding the enzymatic biosynthesis of the metal clusters in Fe_2 ferredoxins, since trisulfides seem to be present in biological systems.

The complexes $\{CpFe(CO)_2\}_2S_n$ have been prepared by several routes (45) (at low temperatures to avoid redox reactions):

$$\left.\begin{array}{ll}
[CpFe(CO)_2]^- + SCl_2 \longrightarrow & (n = 3) \\
[CpFe(CO)_2]^- + S_2Cl_2 \longrightarrow & (n = 3, 4) \\
[CpFe(CO)_2]^- + SOCl_2 \longrightarrow & (n = 3) \\
\{CpFe(CO)_2\}_2 + S_8 \longrightarrow & (n = 3) \\
[CpFe(CO)_2Br] + Li_2S_2 \longrightarrow & (n = 2, 3) \\
[CpFe(CO)_2Br] + Li_2S_4 \longrightarrow & (n = 3, 4)
\end{array}\right\} Cp(CO)_2FeS_nFe(CO)_2Cp$$

Although there is a systematic variation of reactants the trisulfido complex is formed in all the reactions.

Some preparations of polysulfido complexes with sulfur-abstracting reagents have been reported. For example, $[Pt(S_5)_3]^{2-}$ (15), reacts with CN^- to yield $[Pt(S_5)_2]^{2-}$ (168, 192). During the course of this reaction, an interesting two-electron transfer occurs, reducing Pt(IV) to Pt(II). The same substrate reacts with Ph_3P in the following way (42, 81):

$$[Pt(S_5)_3]^{2-} + 12Ph_3P \longrightarrow S^{2-} + (Ph_3P)_2Pt(S_4) + 10Ph_3PS$$

Intramolecular redox reactions within metal–sulfido moieties provide an interesting route to S_2^{2-} complexes. An example is the formation of $[Mo_2O_2S_2(S_2)_2]^{2-}$ from $MoO_2S_2^{2-}$ (161). The following redox processes could be involved (see note added in proof).

$$Mo^{VI} \xrightarrow{red} Mo^{IV}$$
$$(Mo^{IV} + Mo^{VI} \longrightarrow 2Mo^V)$$
$$2S^{2-} \xrightarrow{ox} S_2^{2-}$$

V. Reactions of Coordinated Ligands

Electron-transfer and intramolecular redox reactions (related to S_2^{2-} complexes). The redox behavior of S_2^{2-} complexes is of particular interest because it can probably provide a foundation for understanding the course of reactions involved in relevant enzymes and catalysts (especially hydrodesulfurization catalysts). Intramolecular redox reactions related to type Ia S_2^{2-} ligands can be summarized as follows:

$$M^{n+2}\diagup_S^S \underset{2}{\overset{1}{\rightleftarrows}} M^n\diagup_S^S \underset{4}{\overset{3}{\rightleftarrows}} M^{n-2} + |_S^S$$

$$(2S^{2-}) \qquad (S_2^{2-}) \qquad (S_2^0)$$

Examples of step 1 are provided by oxidation of =S and —SR groups (*85, 132, 159, 161*). Step 2 involves reduction of the S—S bond to form two sulfido groups (*172, 173*). An example of step 3 is thermal decomposition of $Cs_2[Mo_2(S_2)_6] \cdot nH_2O$ to give S_2 as the main gaseous product (*67*) (*vide infra*), and examples of step 4 include synthesis of S_2^{2-} complexes from reactions employing elemental sulfur (*vide supra* and note added in proof).

Reactions with nucleophiles (with abstraction of neutral sulfur). A characteristic reaction of S_n^{2-} ligands is abstraction of a sulfur atom by nucleophiles (N) such as PR_3, SO_3^{2-}, SR^-, CN^-, and OH^-, e.g., the reaction (*87, 105, 110, 116, 125, 130*)

$$M^n\diagup_S^S\diagdown M^n \xrightarrow[-NS]{+N} M^n\diagup^S\diagdown M^n$$

The reaction involves transfer of a neutral sulfur atom from the complex (no change in the oxidation state of the metal atoms occurs). The reaction of $[Mo_3S(S_2)_6]^{2-}$ with CN^- [giving $[Mo_3S_4(CN)_9]^{5-}$] is particularly interesting. The bridging S_2^{2-} ligands are converted into bridging sulfido ligands and the terminal S_2^{2-} ligands are replaced by CN^- groups (*130*).

In $[Mo_4(NO)_4(S_2)_4(S_2)_2O]^{2-}$ (8) both end-on bridging S_2^{2-} ligands of type IIb and the type III bridging S_2^{2-} ligands react with CN^- to yield mainly $[Mo_2S_2(NO)_2(CN)_6]^{6-}$ and (some) $[Mo_4S_4(NO)_4(CN)_8]^{8-}$ (*116, 121*).

The strongly distorted $\{Mo_4S_4\}$ cube of the latter species is produced by abstraction of a sulfur atom from each of the four type III ligands. A reasonable mechanism for the reaction of the two type IIb ligands would be the stepwise sequence

$$\text{Mo}^n\!\!\overset{S-S}{\diagdown}\!\!\text{Mo}^n \xrightarrow[-\text{SCN}^-]{+\text{CN}^-} \text{Mo}^n-S \xrightarrow[-\text{SCN}^-]{+\text{CN}^-} \text{Mo}^{n-2}$$

in which a two-electron reduction occurs parallel to the formation of two metal-to-metal bonds within the $\{Mo_4S_4\}$ cube (or one in $\{Mo_2S_2\}$, respectively.) (*116*).

Type III bridging S_2^{2-} ligands are more susceptible to nucleophilic attack with extrusion of a neutral sulfur atom than are type Ia ligands. This is consistent with a generally lower electron density on the S atoms of type III ligands (bonded to two metal atoms) than of type Ia ligands (bonded to one metal atom).

The compound $Cp_2Ti(S_5)$ (**33**) reacts with PPh_3 to yield the binuclear complex $[Cp_2Ti(S_3)]_2$ (with two S_3^{2-} ligands) (*6*) and with PBu_3 to yield the related $[Cp_2Ti(S_2)]_2$ (with two type IIb S_2^{2-} ligands) (*4*) (for further corresponding reactions of S_n^{2-} ligands with $n > 2$, see below).

Oxidation of the ligand by external agents. In $Ir(dppe)_2(S_2)Cl$, for instance, the S_2^{2-} ligand can be oxidized stepwise (on the metal) to form "S_2O" and "S_2O_2" (*165, 166*). Complexes with bridging "S_2O" ligands are also known (*37*) (a more reasonable classification is as $S_xO_y^{2-}$ ligands).

31   **32**   **33**

Particularly interesting is the oxidation of S_n^{2-} ligands ($n > 2$). The compound $[(S_2)Mo(S)(S)_2Mo(S)(S_4)]^{2-}$ can be oxidized to yield $[(S_2)Mo(O)(S)_2Mo(O)(S_3O_2)]^{2-}$ (**31**) (*131*) and $[(S_2)Mo(S)(S)_2\text{-}Mo(S)(S_3O)]^{2-}$ (**32**) (*160*). The former species is obtained in pure form from a solution containing MoS_4^{2-} simply upon exposure to air (*131*).

Thermal decomposition (with generation of neutral sulfur species, e.g., of S_2). The main gaseous product of thermal decomposition at rather low temperatures (100–200°C) of $Cs_2[Mo_2(S_2)_6]\cdot nH_2O$ is the S_2 molecule, which results by reductive elimination. This has been proved by mass spectroscopy and matrix isolation Raman, UV/VIS, and IR spectroscopy (*67, 107*).

Of general interest also is the thermal decomposition of $(NH_4)_2[Mo_3S(S_2)_6]\cdot nH_2O$ with MoS_2 as the final product (*35, 113*). This implies a reaction without a change of the oxidation state of the metal atoms, according to the equation:

$$(NH_4)_2[Mo_3S(S_2)_6]\cdot nH_2O \longrightarrow 2NH_3 + H_2S + 3MoS_2 + 6S^0 + nH_2O$$

Some additional special reactions of general significance should be mentioned here: $Cp_2Ti(S_5)$ (**33**) has been used as a standard reagent (*167*) for the preparation of new cyclo-S_n species by ligand-transfer reactions [$n = 5 + m$ with S_mCl_2 (*164, 169, 177*); $n = 10, 15, 20$ with SO_2Cl_2 (*179*); with Se_2Cl_2 the cyclo-1,2,3-Se_3S_5 is obtained as final product (*86, 178*)]. Intermediates resulting from the cleavage of the {TiS_5} ring with S^{2-} can be trapped with aliphatic ketones or halides. For example, the isomers $Cp_2Ti(S_2)_2CH_2$ and $Cp_2TiS(S_3)CH_2$ have been obtained from the reaction of **33** with dibromomethane and ammonium sulfide (*56*).

The reactions of polysulfido complexes with activated acetylenes have been studied in some detail (*4, 5*). 1,4-$(MeCp_2Ti)_2(S_2)_2$ reacts with dimethylacetylene dicarboxylate (DMAD) to yield $MeCp_2TiS_2C_2R_2$ with a five membered Ti—S—C=C—S ring as the final product (*4*). Similarly, insertions have also been observed during the reaction of $[Mo_2O_2(S)_2(S_2)_2]^{2-}$ (**3**) with DMAD (*59*). $[MoS(S_4)_2]^{2-}$ (**30a**) reacts with DMAD to yield $[Mo(S_2C_2(COOMe)_2)_3]^{2-}$ (*39*).

Reactions have been observed in which S^{2-}, S_2^{2-}, and S_4^{2-} are formed from each other (particularly in Mo chemistry; cf. ref. (*108*). The formal oxidation state of the metal may also change. With sulfur, MoS_4^{2-} yields $[MoS(S_4)_2]^{2-}$ (**30a**) (*40, 175*), which reacts again with thiolates to form MoS_4^{2-} (*181*). With Ph_2S_2, MoS_4^{2-} forms $[(S_2)MoS(\mu\text{-}S)_2MoS(S_2)]^{2-}$ (*153*) [which inserts (reversibly) CS_2 into the S_2^{2-} groups (*38*) to form a five-membered Mo—S—(C=S)S—S ring]. Upon heating, $MeCp_2V(S_5)$ forms the dinuclear species $MeCpV(\mu\text{-}S)(\mu\text{-}S_2)_2VCpMe$ (*3*).

A transformation reaction occurs when $Cp_2Ti(S_5)$ (**33**) is refluxed for one day. The formation of a second structural isomer (**34**) provides an interesting example for the migration of a π-complexed organic unit to an inorganic sulfur ligand (*57*).

34

VI. Spectroscopic Properties

A. Disulfido Complexes

The $v(S-S)$ vibrational frequencies range from ~ 480 to $600\ cm^{-1}$. Comparison of the $v(S-S)$ values for the discrete diatomic sulfur species S_2 ($^3\Sigma_g^-$: 725 cm^{-1}) (*182*), S_2^- ($^2\Pi_g$: 589 cm^{-1}) (*24, 66*), and S_2^{2-} ($^1\Sigma_g^+$: 446 cm^{-1}) (*67*) leads to the conclusion that the approximate charge distribution is somewhere between that for S_2^- and that for S_2^{2-}. However, here the strong coupling of the $v(S-S)$ vibration with the $v(M-S)$ vibrations, which lead to higher $v(S-S)$ values, must also be considered. This coupling is proved by the shift of 1–2 cm^{-1} observed in some $v(S-S)$ stretching vibrations upon substitution of ^{92}Mo by ^{100}Mo in $[Mo_2(S_2)_6]^{2-}$ and $[Mo_3S(S_2)_6]^{2-}$ (*67*).

Frequencies of type IIIa bridging disulfido ligands are generally higher than those for species with type I structures. In the case of $[Mo_3S(S_2)_6]^{2-}$ with both types of S_2^{2-} ligands the vibrational band at 545 cm^{-1} can be assigned to type IIIa ligands. The higher frequencies found for type IIIa ligands are consistent with the slightly shorter S—S distances for this structural type (see Table II). The intensities of the $v(S-S)$ bands of type IIIa ligands are normally high in the IR as well as in the Raman spectra. The type Ia ligands show intensive $v(S-S)$ bands in the IR, but in the Raman spectrum these bands are usually weak (*67*).

For structure type IIa ligands both the $v(S-S)$ and the totally symmetric $v(M-S)$ vibrations are practically forbidden in the infrared spectrum, but in the Raman spectrum the corresponding bands (intense and strongly polarized) can easily be observed.

X-Ray photoelectron spectra (XPS) have been measured in order to obtain additional information about the effective charge on the sulfur atoms in these ligands. The sulfur $2p$ binding energies lie between ~ 162.9 and 164.4 eV, indicating that the sulfur atoms here are generally more negatively charged than in neutral sulfur [E_B ($2p$) for S_8 is 164.2 eV] (*19*). The corresponding binding energy for sulfur in complexes with reduced sulfur-containing ligands such as S^{2-}, —C—S$^-$, —C—S—C—, or =C=S are in the range 161.5–163.5 eV; in Na$_2$S it has a binding energy of 162.0 eV (*19*), and thiometallates show binding energies of 162.2–163.4 eV (*119*). Thus, the S $2p$ binding energies for disulfido complexes are consistent with the conclusion drawn from S—S distances and from vibrational spectroscopy, i.e., the effective charge on the sulfur atoms in the disulfido ligand in metal complexes is between 0 and -2.

Additional evidence about the oxidation state of the sulfur can be obtained by comparing the metal-binding energies in S_2^{2-} complexes with the electron-binding energies of the metal atom in complexes with known oxidation states. In particular it has become clear that the Mo-binding energies of $[Mo_2(S_2)_6]^{2-}$ and $[Mo_3S(S_2)_6]^{2-}$ can be understood if the ligands are formulated as S_2^{2-} units rather than neutral S_2 fragments (122).[2] Similar results apply for the metal-binding energies of iridium and osmium complexes (47, 72). However, neither the sulfur nor the metal XPS data are sufficiently sensitive to distinguish between the various modes of coordination of the S_2^{2-} ligands.

For bonding type Ia, the π^* orbital of S_2^{2-} splits into a strongly interacting π^*_h orbital in the MS_2 plane and a less-interacting π^*_v orbital perpendicular to the MS_2 plane (cf. Section VII). The longest wavelength band in the electronic spectra of the complexes $Cp_2Nb(S_2)X$ (X = Cl, Br, I) (184) and $MoO(S_2)(dtc)_2$ (98) occurs at ~20 kK and is assigned to ligand to metal charge transfer (LMCT) of the type $\pi^*_v(S) \rightarrow d(M)$. This assignment probably also applies to the corresponding bands of the other complexes of type Ia which contain metal atoms in a high oxidation state. The position of this first band is influenced by the oxidation state of the metal, by the kind of metal–metal bonding, and by the nature of the other ligands, which determine the energy of the LUMO and its metal character.

For S_2^{2-} complexes of type III, the corresponding absorption band is expected to occur at higher energy because both π^* orbitals of the ligand interact strongly with the metal atoms. Particularly interesting is the very intense band at 14.2 kK in $[(NH_3)_5Ru(S_2)Ru(NH_3)_5]Cl_4$. A comparable band is not found in the related compounds of type IIa structure [e.g., $[(CN)_5Co(S_2)Co(CN)_5]^{6-}$ and aqueous $\{Cr(S_2)Cr\}^{4+}$]. It has been proposed that the central unit in $[(NH_3)_5Ru(S_2)Ru(NH_3)_5]^{4+}$ is best formulated as $Ru^{II}-(S_2)^--Ru^{III}$ (15, 44). Such a mixed-valence complex could exhibit the intense band observed (106).

B. S_n^{2-} ($n > 2$) COMPLEXES

The interpretation of vibrational and electronic spectra of these complexes is much more complicated. Whereas IR spectra generally show only bands of low intensity, the Raman spectra often display intense and characteristic lines. Rigorous assignments have not been published so far, but the most simple $M(\mu_2\text{-}S)_n$ ring systems can easily be

---

[2] In the older literature, *disulfur* ligands were often regarded as being neutral (see, e.g., 55, 184).

recognized from the Raman data (the highest wavenumber line is shifted to higher energies with increasing n).

The color and the corresponding electronic spectra are often dominated by ligand internal electronic transitions, the energies of which show a red shift with increasing n. The average negative charge on the sulfur atoms of the S_n^{2-} ligands, of course, decreases with increasing n, as can be seen from XPS data (*107*).

NMR spectroscopy on polysulfido complexes has mainly been used for conformational studies. It has been found that in solution the chair conformation of the TiS_5 cycle in $Cp_2Ti(S_5)$ (**33**) is retained. The ring inversion barrier has been determined for the three $Cp_2M(S_5)$ species with M = Ti, Zr, and Hf (76.3, 48.6, and 58.0 kJ/mol, respectively) (*97*).

VII. Some Structural Features and Chemical Bonding

A. S_2^{2-} COMPLEXES

Structural diversity is achieved through the use of nonbonded pairs of electrons on the S_2^{2-} ligand of both type II complexes to coordinate additional metal atoms. The S—S distances of known complexes range from ~1.98 to 2.15 Å. Most S—S distances are intermediate between the distance of 1.89 Å for S_2 ($^3\Sigma_g^-$) (*104*) and 2.13 Å for S_2^{2-} ($^1\Sigma_g^+$) in Na_2S_2 (*50*). The main S—S distances show no clear systematic trend with structural type (cf. Table II).

The average type Ia S—S distances in $[Mo_2(S_2)_6]^{2-}$ (*128*) and $[Mo_3S(S_2)_6]^{2-}$ (*129*) are slightly longer than the average type IIIa S—S distances. The data for $[Mo_4(NO)_4(S_2)_6O]^{2-}$ (*115*) indicate that type IIb S—S distances may be slightly longer than type IIIa distances. However, caution must be exercised in interpreting these trends because of the known tendency for type IIIa ligands to be slightly asymmetrically bound to the metal atoms.

High coordination numbers of the metal atom are favored for type I and type IIIa structures by the small coordination angles of the bidentate S_2^{2-} ligand. High coordination numbers also protect the metal center from nucleophilic attack, an important factor for the stabilization of metal–metal bonds.

The physical data for dioxygen complexes have been discussed in detail elsewhere (*88, 186*). We note here only that the complexes have been divided into superoxide (O_2^-) and peroxide (O_2^{2-}) ones, primarily on the basis of the O—O distance and $\nu(O-O)$ frequencies: The coordinated superoxide ligand has $d(O-O) \approx 1.30$ Å and

TABLE III

BOND DISTANCES AND VIBRATIONAL
FREQUENCIES FOR X_2^{n-}

	X_2 ($^3\Sigma_g^-$)	X_2^- ($^2\Pi_g$)	X_2^{2-} ($^1\Sigma_g^+$)
d(O—O) (Å)	1.21[a]	1.33[a]	1.49[a]
d(S—S) (Å)	1.89[b]	2.00[c]	2.13[d]
v(O—O) (cm^{-1})	1580[a]	1097[a]	802[a]
v(S—S) (cm^{-1})	725[c]	589[e]	446[f]

[a] Ref. 186.
[b] Ref. 96, 104.
[c] Estimated, see ref. 24.
[d] Ref. 50.
[e] Ref. 24.
[f] Ref. 67.

v(O–O) $\approx$ 1125 cm^{-1}; the coordinated peroxide ligand has d(O–O) $\approx$ 1.45 Å and v(O–O) $\approx$ 860 cm^{-1} (186).

Comparison of the S—S distances and frequencies in Table III indicates that the effective charge on the ligand in the complexes quoted in Table II is between -1 and -2. The S_2^- ligand seems to be much less abundant compared to the corresponding situation with dioxygen complexes. The S_2^- classification has been advocated for the ligand in [(NH$_3$)$_5$Ru(S$_2$)Ru(NH$_3$)$_5$]$^{4+}$ from analysis of the electronic spectra (15, 44) and for Cp$_2$Fe$_2$(S$_2$)(SR)$_2$ (83, 188) from similarities of the ligand geometry to the geometry of the O$_2$ ligand in superoxide complexes.

Relevant to our case is the analysis of the bonding in η^2 side-on S_2^{2-} complexes. Figure 3 shows that the principal bonding interaction is produced by the metal d_{xz} orbital and the π^*_z (π^*_h) orbital of the S_2^{2-} ligand (π-bonding) and also by the d_{z^2} of the metal and the π_z (π_h) (σ-bonding) orbitals. [According to the qualitative Dewar–Chatt–Duncanson bonding scheme (20) the interaction of the π_y (π_v) and the π^*_y (π^*_v) orbitals is considered to be negligible]. The major contribution is expected to be the π-bonding as implied by more quantitative molecular orbital calculation of MO$_2$ complexes with side-on bonded ligands (162).

For type Ia S_2^{2-} complexes, π-bonding should also be the major bonding interaction. For type IIIa M1—(S$_2$)—M2 complexes this bonding interaction would occur twice, once between π^*_z of S_2^{2-} and M1 and once between π^*_y of S_2^{2-} and M2. If M2 is a positively charged

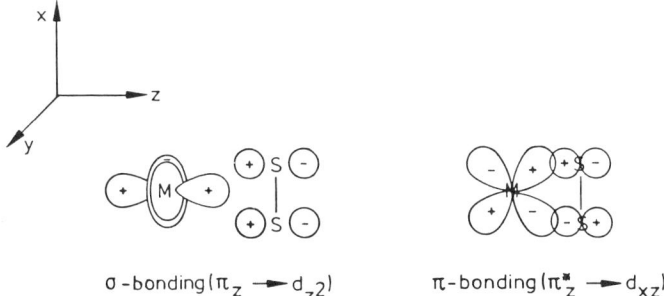

FIG. 3. Schematic bonding interactions between the metal d_{xz} orbital and the π^*_z orbital of the S_2^{2-} ligand (π-bonding) (right) and the d_{z^2} orbital of the metal with the π_z orbital of S_2^{2-} (σ-bonding) (left).

metal ion with few d-electrons, then there should be less electron density on the disulfur unit in a type IIIa complex relative to a type Ia complex. Such a depopulation of the π^* orbitals of the S_2^{2-} ligand is consistent with the greater susceptibility of type IIIa complexes to nucleophilic attack and with the shorter S—S distances and higher ν(S–S) vibrational frequencies for such complexes. These properties are consistent with the π-donor character of the S_2^{2-} ligand.

B. S_n^{2-} ($n > 2$) COMPLEXES

Whereas the coordination of S_2^{2-} ligands to metal atoms always results in a significant shortening of the S—S bond length, the corresponding situation is much more complicated in the case of the S_n^{2-} ($n > 2$) complexes. In general, the shortening of the S—S bond is less pronounced in this type of complex. A semiempirical MO calculation on [Pt(S$_4$)(PH$_3$)$_2$] indicates a substantial donation of electron density from the S$_4$ ligand to the {Pt(PH$_3$)$_2$} fragment, leading to a charge distribution for the coordinated ligand intermediate between those for S$_4$ and S$_4^{2-}$ (12).

Table IV compares bond lengths for S—S bonds in compounds with S$_4$ structural units (uncoordinated and coordinated). In several S_4^{2-} complexes (in addition to a shortening of the S—S bonds compared to the length in S_4^{2-} units in BaS$_4\cdot$H$_2$O), variations of the S—S bond length in the S_4^{2-} chelate are observed. It has been postulated that the high oxidation state of Mo and W in the corresponding complexes contributes to ligand-to-metal π-backbonding and to a shortening of the central S(2)—S(3) bond of the [—S(1)—S(2)—S(3)—S(4)—]$^{2-}$ ligand.

The different conformations occurring in MS$_4$ ring systems have been explained by the varying degree of interligand interactions. In the

TABLE IV

BOND LENGTHS FOR S—S BONDS IN COMPLEXES WITH BIDENTATE S_4^{2-} LIGANDS[a,b,c]

	$d[S(1)-S(2)]$ (Å)	$d[S(2)-S(3)]$ (Å)	$d[S(3)-S(4)]$ (Å)
S_4 Ligand in Complex			
$(Ph_3P)_2Pt(S_4)$	2.024(8)	2.022(10)	2.081(10)
$[PPh_4]_2[Hg(S_4)_2]$	2.050	2.043	2.048
$[Et_4N]_2[Ni(S_4)_2]$	2.073(2)	2.037(4)	2.073(2)
$[Et_4N]_2[Pd(S_4)_2]$	2.062(8)	2.054(6)	2.065(8)
$[AsPh_4]_2[Mo_2S_2(S_4)_2]$[d]	2.019(5)	1.970(6)	2.115(5)
$[AsPh_4]_2[Mo_2S_2(S_4)_2]$[d]	2.096(16)	1.936(19)	2.169(14)
$[PPh_4]_2[Mo_2O_2S_2(S_4)_2]$	2.066	2.024	2.084
$[PPh_4]_2[W_2S_4(S_4)_2]\cdot 0.5DMF$	2.044(11)	2.013(14)	2.112(12)
$[PPh_4]_2[W_2O_2S_2(S_4)_2]\cdot 0.5DMF$	2.062	2.011	2.100
$[PPh_4]_2[W_2S_4(S_2)(S_4)]\cdot 0.5DMF$	2.039	2.016	2.067
$[Et_4N]_2[MoS(S_4)_2]$	2.107(1)	2.012(1)	2.166(1)
$(\eta^5\text{-}C_5H_5)_2Mo(S_4)$	2.081(8)	2.018(9)	2.085(7)
$(\eta^5\text{-}C_5H_5)_2W(S_4)$	2.105(7)	2.016(8)	2.116(9)
Free S_4 Ligand in Salt			
$BaS_4\cdot H_2O$	2.079(3)	2.062(4)	2.079(3)
$C_6H_2(OEt)_2(S_4)_2$[d]	2.028(5)	2.068(5)	2.027(5)
$C_6H_2(OEt)_2(S_4)_2$[d]	2.034(5)	2.067(5)	2.024(5)

[a] Ligand notation is $\overline{S1-S2-S3-S4}$—M.
[b] Compared to the corresponding values of the free ligand in different salts.
[c] Data from refs. *42, 139, 142*.
[d] District S_4 chains are involved..

$Cp_2M(S_4)$ complexes with a tetrahedral coordination of the metal atoms and half-chair conformation of the MS_4 rings these interactions are insignificant (*25, 40*).

In molybdenum and tungsten complexes of the type $[M_2S_2X_2(S_4)(S_n)]^{2-}$ and $[XMo(S_4)_2]^{2-}$ (X = O, S; n = 2, 4) the envelope conformation for the five-membered rings and the obligatory orientation of the lone pairs on the molybdenum-bound sulfur atoms result in a structure with S—S interligand interactions within the XMS_4 pyramidal units being minimized. In these units the half-chair conformation would bring the lone pairs of the sulfur atoms bound to Mo in a position with closer interligand lone-pair contacts (*25, 40*).

For the six-membered MS_5 ring system the preferred conformation is that of a half-chair. The comparison of the molecular parameters of hetero atomic ring systems with those of S_6 in rhombohedral cyclohexasulfur is particularly interesting. The S_6 molecule also has a chairlike structure with an S—S bond length of 2.057(18) Å, an S—S—S bond

TABLE V

Conformations of Most Common Metal–Sulfur Rings

Type	Conformation	Structure	Example formula
MS_4	Half chair (C_2)		$[Ni(S_4)_2]^{2-}$ (**14**)
	Envelope (C_S)		$[Cu_4(S_4)_3]^{2-}$ (**27a**)
MS_5	Chair		$[Cp_2Ti(S_5)]$ (**33**)
MS_6	Chair		$[Ag_2(S_6)_2]^{2-}$ (**28**)
	Twist chair		$[Hg(S_6)_2]^{2-}$ (**17a**)
MS_7	Boat chair		$[Bi_2(S_6)(S_7)_4]^{4-}$ (**25a**)
MS_9			$[Ag(S_9)]^-$ (**18**)
M_2S_3	Envelope		$[Re_4S_{22}]^{4-}$ (**29a**)
$M_2S'S_5$	Crown		$[Mo_2(S_5)(S_2)_3(OH)(NO)_2]^{3-}$ (**20a**)
	Boat chair		$[Nb_2(S_5)(S_2)_3O(OCH_3)_2]^{2-}$ (**20b**)
M_2S_8			$[Au_2S_8]^{2-}$ (**23**)

angle of 102.2(1.6)°, and a torsional angle of ±74.5(2.5)°. Formal replacement of a sulfur atom by {TiCp$_2$} or {VCp$_2$} fragments, for instance, leads to longer M—S distances (2.40–2.46 Å) (compared to the S—S bond length). These are compensated by smaller S—M—S bond angles (89–95°). Therefore, the geometry of the resulting S$_5$ ligand is similar to that of the S$_6$ molecule (*150*).

Conformations of some simple ring systems are summarized in Table V. In some cases different ring systems are observed for the same metal in the same oxidation state [e.g., in [Hg(S$_n$)$_2$]$^{2-}$ (n = 4, 6) (**17a,b**)].

The metal–ligand interaction in S$_n^{2-}$ complexes should be comparable to that in other complexes with sulfur-containing ligands, at least for higher values of n. Additionally, the observation that polysulfides with even n prefer coordination to closed-shell metal ions and those with odd n to open-shell ones indicates that the kind of metal-to-ligand interaction is obviously not restricted to the sulfur atoms attached directly to the metal.

It is interesting to note that (NEt$_4$)$_2$[Ni(S$_4$)$_2$] exists in two polymorphic forms, which differ significantly in their Ni—S distances [d^4(NiS) = 2.185 Å (site symmetry D$_2$) and d^2(NiS) = 2.143 Å and trans to these bonds 2.270 Å (site symmetry of Ni is C$_2$), respectively). Both forms have also been identified and characterized by spectroscopy (*147*).

NOTE ADDED IN PROOF. According to new studies we have to distinguish between *balanced* intramolecular redox processes (*194*) and *unbalanced* ones induced by external reductants or oxidants (*108, 195*).

ACKNOWLEDGMENTS

Our own work reported in this article has been supported by grants from the Deutsche Forschungsgemeinschaft, the Fonds der Chemischen Industrie, and the Minister für Wissenschaft und Forschung, North Rhine-Westphalia. We gratefully acknowledge the contributions of several colleagues and co-workers to these results. We particularly thank Dr. H. Bögge and Dr. K. Schmitz for helpful discussions and for reviewing the manuscript.

REFERENCES

1. Amaudrut, J., Guerchais, J. E., and Sala-Pala, J., *J. Organomet. Chem.* **157,** C 10 (1978).
2. Bird, P. H., McCall, J. M., Shaver, A., and Sirwardane, U., *Angew. Chem.* **94,** 375 (1982); *Angew. Chem. Int. Ed. Engl.* **21,** 384 (1982).
3. Bolinger, C. M., Rauchfuss, T. B., and Rheingold, A. L., *Organometallics* **1,** 1551 (1982).

4. Bolinger, C. M., Hoots, J. E., and Rauchfuss, T. B., *Organometallics* **1,** 223 (1982).
5. Bolinger, C. M., and Rauchfuss, T. B., *Inorg. Chem.* **21,** 3947 (1982).
6. Bolinger, C. M., Rauchfuss, T. B., and Wilson, S. R., *J. Am. Chem. Soc.* **103,** 5620 (1981).
7. Bonds, W. D., Jr., and Ibers, J. A., *J. Am. Chem. Soc.* **94,** 3413 (1972).
8. Böttcher, P., Buchkremer-Hermanns, H., and Baron, J., *Z. Naturforsch. B Anorg. Chem. Org. Chem.* **39b,** 416 (1984).
9. Böttcher, P., and Keller, R., *Z. Naturforsch. B Anorg. Chem. Org. Chem.* **39b,** 577 (1984).
10. Böttcher, P., and Kruse, K., *J. Less-Common Met.* **83,** 115 (1982).
11. Bravard, D. C., Newton, W. E., Huneke, J. T., Yamanouchi, K., and Enemark, J. H., *Inorg. Chem.* **21,** 3795 (1982).
12. Briant, C. E., Calhorda, M. J., Hor, T. S. A., Howells, N. D., and Mingos, D. M. P., *J. Chem. Soc., Dalton Trans.* p. 1325 (1983).
13. Bronger, W., *Angew. Chem.* **93,** 12 (1981); *Angew. Chem. Int. Ed. Engl.* **20,** 52 (1981).
14. Bronger, W., and Spangenberg, M., *J. Less-Common Met.* **76,** 73 (1980).
15. Brulet, C. R., Isied, S. S., and Taube, H., *J. Am. Chem. Soc.* **95,** 4758 (1973).
16. Brunner, H., Wachter, J., Guggolz, E., and Ziegler, M. L., *J. Am. Chem. Soc.* **104,** 1765 (1982).
17. Bunker, M. J., and Green, M. L. H., *J. Chem. Soc., Dalton Trans.*, p. 847 (1981).
18. Burschka, C., *Z. Naturforsch. B Anorg. Chem. Org. Chem.* **35b,** 1511 (1980).
19. Carlson, T. A., "Photoelectron and Auger Spectroscopy." Plenum, New York, 1975.
20. Chatt, J., and Duncanson, L. A., *J. Chem. Soc.*, p. 2939 (1953).
21. Chatt, J., and Mingos, D. M. P., *J. Chem. Soc., A*, p. 1243 (1970).
22. Chen, S., and Robinson, W. R., *J. Chem. Soc., Chem. Commun.*, p. 879 (1978).
23. Clark, G. R., Russell, D. R., Roper, W. R., and Walker, A., *J. Organomet. Chem.* **136,** C 1 (1977).
24. Clark, R. J. H., and Cobbold, D. G., *Inorg. Chem.* **17,** 3169 (1978).
25. Clegg, W., Christou, G., Garner, C. D., and Sheldrick, G. M., *Inorg. Chem.* **20,** 1562 (1981).
26. Clegg, W., Mohan, N., Müller, A., Neumann, A., Rittner, W., and Sheldrick, G. M., *Inorg. Chem.* **19,** 2066 (1980).
27. Clegg, W., Sheldrick, G. M., Garner, C. D., and Christou, G., *Acta Crystallogr. Sect. B* **36,** 2784 (1980).
28. Cloke, P. L., *Geochim. Cosmochim. Acta* **27,** 1265 (1963).
29. Cloke, P. L., *Geochim. Cosmochim. Acta* **27,** 1299 (1963).
30. Coucouvanis, D., Baenziger, N. C., Simhon, E. D., Stremple, P., Swenson, D., Kostikas, A., Simopoulos, A., Petrouleas, V., and Papaefthymiou, V., *J. Am. Chem. Soc.* **102,** 1730 (1980).
31. Coucouvanis, D., Patil, P. R., Kanatzidis, M. G., Detering, B., and Baenziger, N. C., *Inorg. Chem.* **24,** 24 (1985).
32. Coucouvanis, D., Stremple, P., Simhon, E. D., Swenson, D., Baenziger, N. C., Draganjac, M., Chan, L. T., Simopoulos, A., Papaefthymiou, V., Kostikas, A., and Petrouleas, V., *Inorg. Chem.* **22,** 293 (1983).
33. Coucouvanis, D., Swenson, D., Stremple, P., and Baenziger, N. C., *J. Am. Chem. Soc.* **101,** 3392 (1979).
34. Davis, B. R., Bernal, I., and Köpf, H., *Angew. Chem.* **83,** 1018 (1971); *Angew. Chem. Int. Ed. Engl.* **10,** 921 (1971).
35. Diemann, E., Müller, A., and Aymonino, P. J., *Z. Anorg. Allg. Chem.* **479,** 191 (1981).

36. Dirand, J., Ricard, L., and Weiss, R., *Inorg. Nucl. Chem. Lett.* **11**, 661 (1975).
37. Dirand-Colin, J., Schappacher, M., Ricard, L., and Weiss, R., *J. Less-Common Met.* **54**, 91 (1977).
38. Draganjac, M., Ph.D. Thesis, University of Iowa, Ames, 1983.
39. Draganjac, M., and Coucouvanis, D., *J. Am. Chem. Soc.* **105**, 139 (1983).
40. Draganjac, M., Simhon, E. D., Chan, L. T., Kanatzidis, M., Baenziger, N. C., and Coucouvanis, D., *Inorg. Chem.* **21**, 3321 (1982).
41. Drew, M. G. B., Baba, I. B., Rice, D. A., and Williams, D. M., *Inorg. Chim. Acta* **44**, L217 (1980).
42. Dudis, D., and Fackler, J. P., Jr., *Inorg. Chem.* **21**, 3577 (1982).
43. Egen, N. B., and Krause, R. A., *J. Inorg. Nucl. Chem.* **31**, 127 (1969).
44. Elder, R. C., and Trkula, M., *Inorg. Chem.* **16**, 1048 (1977).
45. El-Hinnawi, M. A., Aruffo, A. A., Santarsiero, B. D., McAlister, R., McAlister, D. Q., and Schomaker, V., *Inorg. Chem.* **22**, 1585 (1983).
46. Ellermann, J., Hohenberger, E. F., Kehr, W., Pürzer, A., and Thiele, G., *Z. Anorg. Allg. Chem.* **464**, 45 (1980).
47. Farrar, D. H., Grundy, K. R., Payne, N. C., Roper, W. R., and Walker, A., *J. Am. Chem. Soc.* **101**, 6577 (1979).
48. Fenner, J., Rabenau, A., and Trageser, G., *Adv. Inorg. Chem. Radiochem.* **23**, 329 (1980).
49. Fiechter, S., Kuhs, W. F., and Nitsche, R., *Acta Crystallogr. Sect. B* **36**, 2217 (1980).
50. Föppl, H., Busmann, E., and Frorath, F.-K., *Z. Anorg. Allg. Chem.* **314**, 12 (1962).
51. Furman, N. H., (Ed.), "Standard Methods of Chemical Analysis," 6th Ed., Vol. I, p. 398. Krieger, Huntington, NY, 1975.
52. Giannotti, C., Ducourant, A. M., Chanaud, H., Chiaroni, A., and Riche, C., *J. Organomet. Chem.* **140**, 289 (1977).
53. Gillard, R. D., and Wimmer, F. L., *J. Chem. Soc., Chem. Commun.* p. 936 (1978).
54. Gillard, R. D., Wimmer, F. L., and Richards, J. P. G., *J. Chem. Soc., Dalton Trans.*, p. 253 (1985).
55. Ginsberg, A. P., and Lindsell, W. E., *J. Chem. Soc., Chem. Commun.*, p. 232 (1971).
56. Giolando, D. M., and Rauchfuss, T. B., *Organometallics* **3**, 487 (1984).
57. Giolando, D. M., and Rauchfuss, T. B., *J. Am. Chem. Soc.* **106**, 6455 (1984).
58. Gotzig, J., Rheingold, A. L., and Werner, H., *Angew. Chem.* **96**, 813 (1984); *Angew. Chem. Int. Ed. Engl.* **23**, 814 (1984).
59. Halbert, T. R., Pan, W.-H., and Stiefel, E. I., *J. Am. Chem. Soc.* **105**, 5476 (1983).
60. Haradem, P. S., Cronin, J. L., Krause, R. A., and Katz, L., *Inorg. Chim. Acta* **25**, 173 (1977).
61. Herberhold, M., Reiner, D., Ackermann, K., Thewalt, U., and Debaerdemaeker, T., *Z. Naturforsch. B. Anorg. Chem. Org. Chem.* **39b**, 1199 (1984).
62. Herberhold, M., Reiner, D., Zimmer-Gasser, B., and Schubert, U., *Z. Naturforsch. B. Anorg. Chem. Org. Chem.* **35b**, 1281 (1980).
63. Hieber, W., and Gruber, J., *Z. Anorg. Allg. Chem.* **296**, 91 (1958).
64. Hofmann, K. A., and Höchtlen, F., *Ber. Dtsch. Chem. Ges.* **36**, 3090 (1903).
65. Hofmann, K. A., and Höchtlen, F., *Ber. Dtsch. Chem. Ges.* **37**, 245 (1904).
66. Holzer, W., Murphy, W. F., and Bernstein, H. J., *J. Mol. Spectrosc.* **32**, 13 (1969).
67. Jaegermann, W., Ph.D. Thesis, University of Bielefeld, Federal Republic of Germany, 1981.
68. Jones, P. E., and Katz, L., *J. Chem. Soc., Chem. Commun.*, p. 842 (1967).
69. Jones, P. E., and Katz, L., *Acta Crystallogr. Sect. B* **25**, 745 (1969).
70. Kanatzidis, M. G., Baenziger, N. C., and Coucouvanis, D., *Inorg. Chem.* **22**, 290 (1983).
71. Kettle, S. F. A., and Stanghellini, P. L., *Inorg. Chem.* **16**, 753 (1977).

72. Knecht, J., and Schmid, G., *Z. Naturforsch. B. Anorg. Chem. Org. Chem.* **32b,** 653 (1977).
73. Köpf, H., *Angew. Chem.* **81,** 332 (1969); *Angew. Chem. Int. Ed. Engl.* **8,** 375 (1969).
74. Köpf, H., *Chem. Ber.* **102,** 1509 (1969).
75. Köpf, H., and Block, B., *Chem. Ber.* **102,** 1504 (1969).
76. Köpf, H., Block, B., and Schmidt, M., *Chem. Ber.* **101,** 272 (1968).
77. Köpf, H., Hazari, S. K. S., and Leitner, M., *Z. Naturforsch. B. Anorg. Chem. Org. Chem.* **33b,** 1398 (1978).
78. Köpf, H., Wirl, A., and Kahl, W., *Angew. Chem.* **83,** 146 (1971); *Angew. Chem. Int. Ed. Engl.* **10,** 137 (1971).
79. Krause, R. A., *Inorg. Nucl. Chem. Lett.* **7,** 973 (1971).
80. Krauskopf, K. B., "Introduction to Geochemistry," p. 393. McGraw-Hill, New York, 1979.
81. Kreutzer, B., Kreutzer, P., and Beck, W., *Z. Naturforsch. B. Anorg. Chem. Org. Chem.* **27b,** 461 (1972).
82. Kubas, G. J., and Vergamini, P. J., *Inorg. Chem.* **20,** 2667 (1981).
83. Kubas, G. T., Spiro, T. G., and Terzis, A., *J. Am. Chem. Soc.* **95,** 273 (1973).
84. Küllmer, V., Röttinger, E., and Vahrenkamp, H., *J. Chem. Soc., Chem. Commun.*, p. 782 (1977).
85. Küllmer, V., Röttinger, E., and Vahrenkamp, H., *Z. Naturforsch. B. Anorg. Chem. Org. Chem.* **34b,** 224 (1979).
86. Laitinen, R., Rautenberg, N., Steidel, J., and Steudel, R., *Z. Anorg. Allg. Chem.* **486,** 116 (1982).
87. Leonard, K., Plute, K., Haltiwanger, R. C., and Rakowski DuBois, M., *Inorg. Chem.* **18,** 3246 (1979).
88. Lever, A. B. P., and Gray, H. B., *Acc. Chem. Res.* **11,** 348 (1978).
89. Mague, J. T., and Davis, E. J., *Inorg. Chem.* **16,** 131 (1977).
90. Maheu, L. J., and Pignolet, L. H., *J. Am. Chem. Soc.* **102,** 6346 (1980).
91. Marbach, G., and Strähle, J., *Angew. Chem.* **96,** 229 (1984); *Angew. Chem. Int. Ed. Engl.* **23,** 246 (1984).
92. Marcoll, J., Rabenau, A., Mootz, D., and Wunderlich, H., *Rev. Chim. Miner.* **11,** 607 (1974).
93. Markó, L., Bor, G., Klumpp, E., Markó, B., and Almásy, G., *Chem. Ber.* **96,** 955 (1963).
94. Martin, T. P., and Schaber, H., *Spectrochim. Acta Part A* **38A,** 655 (1982).
95. Matthes, S., "Mineralogie, Eine Einführung in die spezielle Mineralogie, Petrologie und Lagerstättenkunde," p. 237. Springer-Verlag, Berlin and New York, 1983.
96. Maxwell, L. R., Mosley, V. M., and Hendricks, S. B., *Phys. Rev.* **50,** 41 (1936).
97. McCall, J. M., and Shaver, A., *J. Organomet. Chem.* **193,** C37 (1980).
98. McDonald, J. W., and Newton, W. E., *Inorg. Chim. Acta* **44,** L81 (1980).
99. Mealli, C., and Midollini, S., *Inorg. Chem.* **22,** 2785 (1983).
100. Mennemann, K., and Mattes, R., *Angew. Chem.* **89,** 269 (1977); *Angew. Chem. Int. Ed. Engl.* **16,** 260 (1977).
101. Mennemann, K., and Mattes, R., *J. Chem. Res. (M)*, p. 1372 (1979).
102. Mercier, R., Douglade, J., Amaudrut, J., Sala-Pala, J., and Guerchais, J., *Acta Crystallogr. Sect. B* **36,** 2986 (1980).
103. Meunier, B., and Prout, K., *Acta Crystallogr. Sect. B* **35,** 172 (1979).
104. Meyer, B., *Chem. Rev.* **76,** 367 (1976).
105. Miller, K. F., Bruce, A. E., Corbin, J. L., Wherland, S., and Stiefel, E. I., *J. Am. Chem. Soc.* **102,** 5102 (1980).
106. Miskowski, V. M., Robbinds, J. L., Treitel, I. M., and Gray, H. B., *Inorg. Chem.* **14,** 2318 (1975).

107. Müller, A., *et al.*, unpublished results.
108. Müller, A., *Polyhedron* **5,** 323 (1986): *Proc. 5th Int. Conf. Molybdenum, Newcastle,* 1985.
109. Müller, A., Baumann, F.-W., Bögge, H., Römer, M., Krickemeyer, E., and Schmitz, K., *Angew. Chem.* **96,** 607 (1984); *Angew. Chem. Int. Ed. Engl.* **23,** 632 (1984).
110. Müller, A., Bhattacharyya, R. G., Mohan, N., and Pfefferkorn, B., *Z. Anorg. Allg. Chem.* **454,** 118 (1979).
111. Müller, A., Bhattacharyya, R. G., and Pfefferkorn, B., *Chem. Ber.* **112,** 778 (1979).
112. Müller, A., Cyrankiewicz, R., and Bögge, H., unpublished results; Cyrankiewicz, R., Diploma thesis, University of Bielefeld, Federal Republic of Germany, 1985.
113. Müller, A., and Diemann, E., *Chimia* **39,** 312 (1985).
114. Müller, A., Eltzner, W., Bögge, H., and Krickemeyer, E., *Angew. Chem.* **95,** 905 (1983); *Angew. Chem. Int. Ed. Engl.* **22,** 884 (1983).
115. Müller, A., Eltzner, W., Bögge, H., and Sarkar, S., *Angew. Chem.* **94,** 555 (1982); *Angew. Chem. Int. Ed. Engl.* **21,** 535 (1982); *Angew. Chem. Suppl.,* p. 1167 (1982).
116. Müller, A., Eltzner, W., Clegg, W., and Sheldrick, G. M., *Angew. Chem.* **94,** 555 (1982); *Angew. Chem. Int. Ed. Engl.* **21,** 536 (1982).
117. Müller, A., Eltzner, W., Jostes, R., Bögge, H., Diemann, E., Schimanski, J., and Lueken, H., *Angew. Chem.* **96,** 355 (1984); *Angew. Chem. Int. Ed. Engl.* **23,** 389 (1984).
118. Müller, A., Eltzner, W., and Mohan, N., *Angew. Chem.* **91,** 158 (1979); *Angew. Chem. Int. Ed. Engl.* **18,** 168 (1979).
119. Müller, A., Jørgensen, C. K., and Diemann, E., *Z. Anorg. Allg. Chem.* **391,** 38 (1972).
120. Müller, A., Jostes, R., and Cotton, F. A., *Angew. Chem.* **92,** 921 (1980); *Angew. Chem. Int. Ed. Engl.* **19,** 875 (1980).
121. Müller, A., Jostes, R., Eltzner, W., Nie, C.-S., Diemann, E., Bögge, H., Zimmermann, M., Dartmann, M., Reinsch-Vogell, U., Che, S., Cyvin, S. J., and Cyvin, B. N., *Inorg. Chem.* **24,** 2872 (1985).
122. Müller, A., Jostes, R., Jaegermann, W., and Bhattacharyya, R. G., *Inorg. Chim. Acta* **41,** 259 (1980).
123. Müller, A., Krickemeyer, E., Bögge, H., Clegg, W., and Sheldrick, G. M., *Angew. Chem.* **95,** 1030 (1983); *Angew. Chem. Int. Ed. Engl.* **22,** 1006 (1983).
124. Müller, A., Krickemeyer, E., and Bögge, H., *Angew. Chem.* **98,** 258 (1986); *Angew. Chem. Int. Ed. Engl.* **25,** 272 (1986).
125. Müller, A., Krickemeyer, E., and Reinsch, U., *Z. Anorg. Allg. Chem.* **470,** 35 (1980).
126. Müller, A., Krickemeyer, E., Zimmermann, M., Römer, M., Bögge, H., Penk, M., and Schmitz, K., *Inorg. Chim. Acta* **90,** L69 (1984).
127. Müller, A., Nolte, W.-O., and Krebs, B., *Angew. Chem.* **90,** 286 (1978); *Angew. Chem. Int. Ed. Engl.* **17,** 279 (1978).
128. Müller, A., Nolte, W.-O., and Krebs, B., *Inorg. Chem.* **19,** 2835 (1980).
129. Müller, A., Pohl, S., Dartmann, M., Cohen, J. P., Bennett, J. M., and Kirchner, R. M., *Z. Naturforsch. B. Anorg. Chem. Org. Chem.* **34b,** 434 (1979).
130. Müller, A., and Reinsch, U., *Angew. Chem.* **92,** 69 (1980); *Angew. Chem. Int. Ed. Engl.* **19,** 72 (1980).
131. Müller, A., Reinsch-Vogell, U., Krickemeyer, E., and Bögge, H., *Angew. Chem.* **94,** 784 (1982); *Angew. Chem. Int. Ed. Engl.* **21,** 796 (1982).
132. Müller, A., Rittner, W., Neumann, A., and Sharma, R. C., *Z. Anorg. Allg. Chem.* **472,** 69 (1981).
133. Müller, A., Römer, M., Bögge, H., Krickemeyer, E., Baumann, F.-W., and Schmitz, K., *Inorg. Chim. Acta* **89,** L7 (1984).

134. Müller, A., Römer, M., Bögge, H., Krickemeyer, E., and Bergmann, D., *J. Chem. Soc., Chem. Commun.*, p. 348 (1984).
135. Müller, A., Römer, M., Bögge, H., Krickemeyer, E., and Bergmann, D., *Z. Anorg. Allg. Chem.* **511,** 84 (1984).
136. Müller, A., Römer, M., Bögge, H., Krickemeyer, E., and Schmitz, K., *Inorg. Chim. Acta* **85,** L39 (1984).
137. Müller, A., Römer, M., Bögge, H., Krickemeyer, E., and Zimmermann, M., *Z. Anorg. Allg. Chem.* **534,** 69 (1986).
138. Müller, A., Römer, M., Krickemeyer, E., and Bögge, H., *Naturwissenschaften* **71,** 43 (1984).
139. Müller, A., Römer, M., Römer, C., Reinsch-Vogell, U., Bögge, H., and Schimanski, U., *Monatsh. Chem.* **116,** 711 (1985).
140. Müller, A., Sarkar, S., Bhattacharyya, R. G., Pohl, S., and Dartmann, M., *Angew. Chem.* **90,** 564 (1978); *Angew. Chem. Int. Ed. Engl.* **17,** 535 (1978).
141. Müller, A., Sarkar, S., Bögge, H., Jostes, R., Trautwein, A., and Lauer, U., *Angew. Chem.* **95,** 574 (1983); *Angew. Chem. Int. Ed. Engl.* **22,** 561 (1983); *Angew. Chem. Suppl.*, p. 747 (1983).
142. Müller, A., Schimanski, J., Schimanski, U., and Bögge, H., *Z. Naturforsch. B. Anorg. Chem. Org. Chem.* **40b,** 1277 (1985).
143. Müller, A., Schimanski, J., Römer, M., Bögge, H., Baumann, F.-W., Eltzner, W., Krickemeyer, E., and Billerbeck, U., *Chimia* **39,** 25 (1985).
144. Müller, A., Schimanski, J., and Schimanski, U., *Angew. Chem.* **96,** 158 (1984); *Angew. Chem. Int. Ed. Engl.* **23,** 159 (1984).
145. Müller, A., Schimanski, J., Schimanski, U., and Bögge, H., *Z. Naturforsch. B. Anorg. Chem. Org. Chem.* **40b,** 1277 (1985).
146. Müller, A., and Schladerbeck, N., *Chimia* **39,** 23 (1985).
147. Müller, A., Schmitz, K., Krickemeyer, E., Römer, M., Bögge, H., and Baumann, F.-W., *J. Mol. Struct.*, in press.
148. Müller, A., Schmitz, K., Krickemeyer, E., Penk, M., and Bögge, H., *Angew. Chem.* **98,** 470 (1986); *Angew. Chem. Int. Ed. Engl.* **25,** 453 (1986).
149. Müller, A., Zimmermann, M., and Bögge, H., *Angew. Chem.* **98,** 259 (1986); *Angew. Chem. Int. Ed. Engl.* **25,** 273 (1986).
150. Muller, E. G., Petersen, J. L., and Dahl, L. F., *J. Organomet. Chem.* **111,** 91 (1976).
151. Nappier, T. E., Jr., and Meek, D. W., *J. Am. Chem. Soc.* **94,** 306 (1972).
152. Nappier, T. E., Jr., Meek, D. W., Kirchner, R. M., and Ibers, J. A., *J. Am. Chem. Soc.* **95,** 4194 (1973).
153. Pan, W.-H., Harmer, M. A., Halbert, T. R., and Stiefel, E. I., *J. Am. Chem. Soc.* **106,** 459 (1984).
154. Piper, T. S., and Wilkinson, G., *J. Am. Chem. Soc.* **78,** 900 (1956).
155. Perrin-Billot, C., Perrin, A., and Prigent, J., *J. Chem. Soc., Chem. Commun.*, p. 676 (1970).
156. Petersen, J. L., and Dahl, L. F., *J. Am. Chem. Soc.* **97,** 6416 (1975).
157. Potvin, C., Brégeault, J.-M., and Manoli, J.-M., *J. Chem. Soc., Chem. Commun.*, p. 664 (1980).
158. Rakowski DuBois, M., DuBois, D. L., VanDerveer, M. C., and Haltiwanger, R. C., *Inorg. Chem.* **20,** 3064 (1981).
159. Ramasami, T., Taylor, R. S., and Sykes, A. G., *J. Chem. Soc., Chem. Commun.*, p. 383 (1976).
160. Reinsch-Vogell, U., Ph.D. Thesis, University of Bielefeld, Federal Republic of Germany, 1984.

161. Rittner, W., Müller, A., Neumann, A., Bäther, W., and Sharma, R. C., *Angew. Chem.* **91,** 565 (1979); *Angew. Chem. Int. Ed. Engl.* **18,** 530 (1979).
162. Sakaki, S., Hori, K., and Ohyoshi, A., *Inorg. Chem.* **17,** 3183 (1978).
163. Samuel, E., and Gianotti, G., *J. Organomet. Chem.* **113,** C17 (1976).
164. Sandow, T., Steidel, J., and Steudel, R., *Angew. Chem.* **94,** 782 (1982); *Angew. Chem. Int. Ed. Engl.* **21,** 788 (1982).
165. Schmid, G., and Ritter, G., *Angew. Chem.* **87,** 673 (1975); *Angew. Chem. Int. Ed. Engl.* **14,** 645 (1975).
166. Schmid, G., Ritter, G., and Debaerdemaeker, T., *Chem. Ber.* **108,** 3008 (1975).
167. Schmidt, M., Block, B., Block, H. D., Köpf, H., and Wilhelm, E., *Angew. Chem.* **80,** 660 (1968); *Angew. Chem. Int. Ed. Engl.* **7,** 632 (1968).
168. Schmidt, M., and Hoffmann, G. G., *Z. Anorg. Allg. Chem.* **452,** 112 (1979).
169. Schmidt, M., and Wilhelm, E., *J. Chem. Soc., Chem. Commun.*, p. 1111 (1970).
170. v. Schnering, H. G., and Beckmann, W., *Z. Anorg. Allg. Chem.* **347,** 231 (1966).
171. Schunn, R. A., Fritchie, C. J., Jr., and Prewitt, C. T., *Inorg. Chem.* **5,** 892 (1966).
172. Seyferth, D., and Henderson, R. S., *J. Am. Chem. Soc.* **101,** 508 (1979).
173. Seyferth, D., Henderson, R. S., and Gallagher, M. K., *J. Organomet. Chem.* **193,** C75 (1980).
174. Siebert, H., and Thym, S., *Z. Anorg. Allg. Chem.* **399,** 107 (1973).
175. Simhon, E. D., Baenziger, N. C., Kanatzidis, M., Draganjac, M., and Coucouvanis, D., *J. Am. Chem. Soc.* **103,** 1218 (1981).
176. Spangenberg, M., and Bronger, W., *Z. Naturforsch. B. Anorg. Chem. Org. Chem.* **33b,** 482 (1978).
177. Steidel, J., and Steudel, R., *J. Chem. Soc., Chem. Commun.*, p. 1312 (1982).
178. Steudel, R., and Strauss, E.-M., *Angew. Chem.* **96,** 356 (1984); *Angew. Chem. Int. Ed. Engl.* **23,** 362 (1984).
179. Steudel, R., and Strauss, R., *J. Chem. Soc., Dalton Trans.*, p. 1775 (1984).
180. Stevenson, D. L., Magnuson, V. R., and Dahl, L. F., *J. Am. Chem. Soc.* **89,** 3727 (1967).
181. Stiefel, E. I., *Proc. Int. Conf. Chem. Uses Molybdenum 4th* (1982).
182. Suchard, S. N., and Melzer, J. E., in "Spectroscopic Constants for Selected Homonuclear Diatomic Molecules," Vol. II. Report SAMSO-TR-76-31, 1976.
183. Terzis, A., and Rivest, R., *Inorg. Chem.* **12,** 2132 (1973).
184. Treichel, P. M., and Werber, G. P., *J. Am. Chem. Soc.* **90,** 1753 (1968).
185. Uchtman, V. A., and Dahl, L. F., *J. Am. Chem. Soc.* **91,** 3756 (1969).
186. Vaska, L., *Acc. Chem. Res.* **9,** 175 (1976).
187. Vergamini, P. J., and Kubas, G. J., *Prog. Inorg. Chem.* **21,** 261 (1976).
188. Vergamini, P. J., Ryan, R. R., and Kubas, G. J., *J. Am. Chem. Soc.* **98,** 1980 (1976).
189. Wakatsuki, Y., and Yamazaki, H., *J. Organomet. Chem.* **64,** 393 (1974).
190. Wei, C. H., and Dahl, L. F., *Inorg. Chem.* **4,** 1 (1965).
191. Wells, A. F., "Structural Inorganic Chemistry," 5th ed., p. 727. Oxford Univ. Press (Clarendon), London, 1984.
192. Wickenden, A. E., and Krause, R. A., *Inorg. Chem.* **8,** 779 (1969).
193. Yamanouchi, K., Huneke, J. T., and Enemark, J. H., in "Molybdenum Chemistry of Biological Significance" (W. E. Newton, and S. Otsuka, Eds.), p. 309. Plenum, New York, 1980.
194. Müller, A., Diemann, E., Wienböker, U., and Bögge, H., *Inorg. Chem.*, in press.
195. Stiefel, E. I., *Polyhedron* **5,** 341 (1986).

IMINOBORANES

PETER PAETZOLD

Institut für Anorganische Chemie der Rheinisch-Westfälischen Technischen Hochschule,
D-5100 Aachen, Federal Republic of Germany

I. Introduction

Iminoboranes may be identified as a class of molecules with an imino group NR and a varying group X (e.g., F, RO, R_2N, R_3C) bonded to boron. Iminoboranes (XBNR) belong to the family of neutral two-coordinated boron species which may be arranged systematically in the following way:

$XBCR_2$	XBNR	XBO
$XBSiR_2$	XBPR	XBS
⋮	⋮	⋮

Experimental data for alkylidenoboranes ($XBCR_2$) (1), oxoboranes (XBO) (2), and thioboranes (XBS) (3–7) are described in the literature. All these molecules exhibit novel multiple bonds between boron and its neighboring atoms. For some years, doubly or triply bonded main group elements, involving not only boron, but also silicon, phosphorus, sulfur, and others, have modified the traditional qualitative concepts of bonding. As a counterpart, chemists have also become aware of multiple bonds, including quadruple bonds, between transition metals.

In particular, iminoboranes (XBNR) are isoelectronic with alkynes (XCCR). Well-known comparable pairs of isoelectronic species are aminoboranes (X_2BNR_2) and alkenes (X_2CCR_2), amine–boranes (X_3BNR_3) and alkanes (X_3CCR_3), borazines [$(XBNR)_3$] and benzenes [$(XCCR)_3$], etc. The structure of aminoboranes, amine–boranes, and borazines is well known from many examples. It has turned out that these BN species are not only isoelectronic, but also have structures comparable with the corresponding CC species. In the case of borazines, the aromatic character was widely discussed on the basis of theoretical and experimental arguments. The structural and physical properties of

BN species parallel those of CC. This parallelism does not hold so nicely for reactivity. The polarity and relative weakness of B—N bonds make BN species much more reactive than comparable CC species, at least with respect to polar additions or substitutions, and the reaction paths as well as the products may differ greatly. The chemistry of BN compounds with three- or four-coordinated B and N atoms is summarized in advanced textbooks, and the details can be found in Gmelin's handbook.

This article describes what is known about the formation, structure, and reactivity of iminoboranes. The chemistry of iminoboranes is in its beginnings, and so we cannot paint a complete picture. A comparison between iminoboranes and the corresponding alkynes will serve as a background throughout this novel field of boron chemistry.

II. Formation of Iminoboranes

A. Metastable Iminoboranes

Since iminoboranes are thermodynamically unstable with respect to their oligomerization (Section IV,A), they can be isolated under laboratory conditions by making the oligomerization kinetically unfavorable. Low temperature, high dilution, bulky groups R and X, or a combination of these is necessary.

The parent compound HBNH was the first iminoborane to be identified unequivocally (8). It was formed by photolysis of solid H_3BNH_3 and trapped, together with further products, in a rare gas matrix which provided low temperature and high dilution for this extremely unshielded molecule. Evidence for the iminoborane in the mixture came from a vibrational analysis, all four atoms in turn being isotopically labeled.

The first iminoborane that could be handled at $-30°C$ in a liquid 1:1 mixture with Me_3SiCl was $(F_5C_6)BNtBu$ (9). It was synthesized by the general elimination reaction in Eq. (1) (Hal = halide) from the corresponding aminoborane. A "hot tube procedure" must be employed for

$$\begin{array}{c} X \\ {}_{Hal}\!\!\diagdown\!\!B\!=\!N\!\!\diagup\!\!{}^{R}_{SiMe_3} \end{array} \xrightarrow{-Me_3SiHal} XBNR \qquad (1)$$

this type of elimination with temperatures near 500°C and pressures near 10^{-3} Torr. The products are trapped at liquid nitrogen temperature. The halosilane can be separated by low-temperature evaporation, provided that the iminoborane formed is relatively stable (10–14). The extreme shielding effect of X = (Me$_3$Si)$_3$C and X = (Me$_3$Si)$_3$Si makes the corresponding iminoboranes (R = SiMe$_3$) storable at room temperature (15); it may be that these particular iminoboranes are stable toward oligomerization even in a thermodynamic sense. The compound (Me$_3$Si)$_3$CBNSiMe$_3$ could be prepared in the liquid phase at 60°C, according to Eq. (1). The 2,2,6,6-tetramethylpiperidino group, C$_9$H$_{18}$N, also exhibits a strong shielding effect, so that the aminoiminoborane C$_9$H$_{18}$NBNtBu may be synthesized at room temperature by base-induced HCl elimination after Eq. (2), without considerable product loss by dimerization (16).

$$\underset{\substack{Cl \quad H}}{\overset{\substack{N \quad tBu \\ B=N}}{\bigcirc}} \quad \xrightarrow[-\text{HNC}_9\text{H}_{18}]{+\text{LiNC}_9\text{H}_{18}} \quad \underset{-\text{LiCl}}{\overset{}{\bigcirc}} \text{NBN} t\text{Bu} \qquad (2)$$

Thermal decomposition of azidoboranes [Eq. (3)], and of [trimethylsilyl(trimethylsilyloxy)amino]boranes [Eq. (4)], permits a simple synthesis of "symmetric" iminoboranes RBNR (17–19). A hot tube procedure at about 300°C and 10^{-3} Torr turned out to be useful. The iminoborane iPrBNiPr, for example, was prepared at a rate of 10 g/hour by the azidoborane method. No separation problems are met with this method. Handling the liquid reactants of Eqs. (3) and (4) is hazardous.

$$R_2BN_3 \xrightarrow{-N_2} RBNR \qquad (3)$$

$$R_2B=N\overset{OSiMe_3}{\underset{SiMe_3}{\diagdown}} \xrightarrow{-(Me_3Si)_2O} RBNR \qquad (4)$$

The violence of uncontrolled decompositions decreases markedly with the bulk of R, and explosions are more violent in the case of azidoboranes. Hot tube experiments at 400°C, aiming at the preparation of aminoiminoboranes XBNR [X = Me$_3$Si(tBu)N, R = iPr, Bu] from the corresponding azidoboranes XRBN$_3$, have been unsuccessful insofar as the cyclodimers of XBNR were the only isolable products (14).

At present, 17 iminoboranes are known as sufficiently well-characterized liquids (Table I). Spectroscopically, there are two typical

TABLE I

Isolated Iminoboranes XBNR

X	R	Preparation equation number	Spectroscopic evidence	Reference
Me	Me	(3), (4)	NMR (^{1}H, ^{11}B, ^{13}C), IR, PE	19
Me	tBu	(1)	NMR (^{1}H, ^{11}B, ^{13}C, ^{14}N), IR	13
Et	Et	(4)	NMR (^{1}H, ^{11}B), IR	19
Et	tBu	(1)	NMR (^{1}H, ^{11}B, ^{13}C, ^{14}N), IR	10
Pr	tBu	(1)	NMR (^{1}H, ^{11}B, ^{13}C, ^{14}N), IR	10
iPr	iPr	(3), (4)	NMR (^{1}H, ^{11}B), IR	17, 19
iPr	tBu	(1)	NMR (^{1}H, ^{11}B, ^{13}C, ^{14}N), IR	12
Bu	tBu	(1)	NMR (^{1}H, ^{11}B, ^{13}C, ^{14}N), IR	10
iBu	iBu	(3)	NMR (^{1}H, ^{11}B), IR	17
sBu	sBu	(3)	NMR (^{1}H, ^{11}B), IR	17
sBu	tBu	(1)	NMR (^{11}B), IR	19
tBu	tBu	(1)	NMR (^{1}H, ^{11}B, ^{13}C, ^{14}N), IR, Raman, PE	11
F$_5$C$_6$	tBu	(1)	NMR (^{1}H, ^{11}B, ^{13}C, ^{19}F), IR	9
(Me$_3$Si)$_3$C	SiMe$_3$	(1)	NMR (^{1}H, ^{11}B, ^{13}C, ^{14}N, ^{29}Si), IR	15
(Me$_3$Si)$_3$Si	SiMe$_3$	(1)	NMR (^{1}H, ^{11}B, ^{13}C, ^{14}N, ^{29}Si), IR	15
(2,2,6,6-tetramethylpiperidino) N	tBu	(2)	NMR (^{1}H, ^{11}B, ^{13}C), IR	16
(tBu)(Me$_3$Si)N	tBu	(1)	NMR (^{1}H, ^{11}B), IR	14

features, easily accessible at low temperature: (1) ^{11}B-NMR chemical shifts are found in a range from 2.3 to 6.3 ppm (Et$_2$O·BF$_3$ as the external standard), depending on the ligands X and R and on the solvent. These shifts are far from the ^{11}B shifts of the corresponding oligomers. Outside of that range, an ^{11}B signal appears at −2.7 ppm with (F$_5$C$_6$)BNtBu, apparently due to the extraordinary electronic effect of the pentafluorophenyl group. ^{11}B-NMR shifts at 21.0 and 21.9 ppm for the iminoboranes with R = SiMe$_3$ are far away from the typical range, perhaps due to the N-bonded SiMe$_3$ group. (2) A broad intense absorption peak in the range 2008–2038 cm^{-1} is observed in the infrared spectrum of alkyl(alkylimino)boranes R'BNR, together with a less intense band in the range 2063–2090 cm^{-1}. These bands can be assigned to the ^{11}BN and ^{10}BN vibrations, respectively (Section III,C). Amino-iminoboranes cause absorptions at lower frequencies: 1990 cm^{-1} for the N^{11}BN and 2020–2025 cm^{-1} for the N^{10}BN asymmetric stretching vibration. The $^{11}BN/^{10}BN$ band couple for $[(Me_3Si)_3E]BNSiMe_3$ falls out of the typical range: 1885, 1940 cm^{-1} (E = C) and 1985, 1925 cm^{-1} (E = Si). A monomeric structure including a linear CBNC skeleton was proven for tBuBNtBu in the solid state at $-85°C$ (Section III,B). Since the typical spectroscopic data for this particular molecule are in accord with the data of all iminoboranes R'BNR, one can ascribe the same structural features to all of them.

Iminoboranes exhibit marked differences in their kinetic stability. The methyl derivative is the most unstable, as expected; cocondensation of equal amounts of MeBNMe, $(Me_3Si)_2O$ [Eq. (4)], and pentane gives a mixture that is liquid at $-90°C$ and needs to be worked up quickly for spectra or further reactions; after 2 days at $-90°C$, MeBNMe is no longer detectable, and the stabilization product, $(MeBNMe)_3$, is a by-product from the beginning. The stability of EtBNEt is not much greater, whereas pure liquid iBuBNiBu with β-branched alkyl groups may be handled, but not stored, at $-80°C$. α-Branched alkyl groups markedly stabilize iminoboranes; iPrBNiPr and sBuBNsBu can be stored as pure liquids at $-80°C$ for some time, but trimerize rapidly at room temperature, liberating a detectable amount of heat. Finally, 1 g of tBuBNtBu with the doubly α-branched tBu groups dimerizes with a half-life of 3 days at $+50°C$; this iminoborane may be handled as a normal, air-sensitive organoborane at $0°C$. The two aminoiminoboranes (Table I) seem to be even more stable.

Production of iminoboranes by a hot tube procedure is obviously restricted to those reactants that can be transported into the gas phase without decomposition. Owing to this restriction the reactions according to Eqs. (3) and (4) cannot be applied to reactants with alkyl groups larger than C_4H_9 or with aryl groups.

B. IMINOBORANES AS INTERMEDIATES

Equation (5) represents a general formation of borazines, performed in solution or in the melt from the starting aminoborane, carried out either thermally or with the aid of bases. Apparently, Eq. (5) is a more or

$$3 \begin{array}{c} X \\ \diagdown \\ B{=}N \\ \diagup \quad \diagdown \\ Y \quad\quad A \end{array} \xrightarrow{-3\ AY} \begin{array}{c} X \\ R\diagdown \overset{|}{B}\diagup R \\ N\quad\quad N \\ |\quad\quad\quad | \\ B\diagdown \quad\diagup B \\ X \quad N \quad X \\ R \end{array} \quad (5)$$

A = H, R'$_2$B, Me$_3$Si, etc.

Y = H, F, Cl, Br, R'O, R'$_2$N, etc.

less complicated multistep process. If we consider the first step only, there will be at least two plausible mechanistic alternatives: (1) formation of iminoborane by elimination of AY, [Eq. (6)], and (2) a bimolecular association equilibrium [Eq. (7)]. We do not know of a proposed mechanism via iminoboranes that is supported by substantial evidence.

$$\underset{Y}{\overset{X}{\diagdown}}B=N\underset{A}{\overset{R}{\diagup}} \xrightarrow{-AY} XBNR \xrightarrow{+ \overset{X}{\underset{Y}{\diagdown}}B=N\underset{A}{\overset{R}{\diagup}}} \underset{Y}{\overset{X}{\diagdown}}B\overset{R}{\underset{|}{-}}N\overset{X}{\underset{|}{-}}B\underset{A}{\overset{R}{\diagup}} \qquad (6)$$

$$\underset{Y}{\overset{X}{\diagdown}}B=N\underset{A}{\overset{R}{\diagup}} + \underset{Y}{\overset{X}{\diagdown}}B=N\underset{A}{\overset{R}{\diagup}} \rightleftarrows \underset{Y}{\overset{X}{\diagdown}}B\underset{\underset{Y}{|}}{\overset{\overset{R}{|}}{-}}N-B\underset{A}{\overset{R}{\diagup}} \xrightarrow{-AY} \underset{Y}{\overset{X}{\diagdown}}B\overset{R}{\underset{|}{-}}N\overset{X}{\underset{|}{-}}B\underset{A}{\overset{R}{\diagup}} \qquad (7)$$

Equation (8) represents the decomposition of liquid azidoboranes at a temperature lower than 100°C (17). At higher temperatures, additional decomposition transforms the products into the corresponding borazines $(XBNR)_3$, according to Eq. (5).

$$2 \, R_2BN_3 \xrightarrow{-N_2} R_2B\overset{..}{-}NR\overset{..}{-}BR\overset{..}{-}N_3 \qquad (8)$$

R = Pr, iPr, Bu, iBu, sBu, n-C_6H_{11}

Iminoboranes were suggested as intermediates, which were azidoborated by an excess of reactants. It was argued that the same products are formed when isolated iminoboranes, RBNR, produced according to Eqs. (1), (3), and (4), are azidoborated with R_2BN_3 (Section V,C). Further support for iminoboranes as intermediates came from the observation that the trapping function of the azidoborane in Eq. (8) can be overcome by an excess of independent trapping agents (e.g., trialkylboranes BR'_3, the reagents that are known to give the same products in reactions with isolated iminoboranes) (10–14, 17, 18). Transposing the mechanistic alternatives of Eqs. (6) and (7) to the decomposition of R_2BN_3 in the presence of BR'_3, we should consider the two final products of Eqs. (9) and (10), which are easily distinguishable by NMR.

$$R_2BN_3 \xrightarrow{-N_2} RBNR \xrightarrow{+ BR'_3} RR'B\overset{..}{-}NR\overset{..}{-}BR'_2 \qquad (9)$$

$$R_2BN_3 \underset{-BR'_3}{\overset{+BR'_3}{\rightleftarrows}} R_2B-N\underset{N_2}{\overset{BR'_3}{\diagup}} \xrightarrow{-N_2} R_2B\dot{-}NR'\dot{-}BR'_2 \qquad (10)$$

With the ligands R of Eq. (8) and with BEt$_3$ as the trapping agent, the diborylamines of Eq. (9) were formed, but no product of Eq. (10) was detected (17), thus strongly supporting the idea of iminoborane intermediates.

The situation is different when diarylazidoboranes, Ar$_2$BN$_3$, decompose. First, the overall reaction differs from Eq. (8). Simple aryl groups like phenyl or o-tolyl cause the formation of iminoborane cyclooligomers, without primary trapping products being isolable. The pentafluorophenyl or the mesityl group makes products available that might have been formed again by trapping iminoboranes by the starting compounds; but instead of azidoboration products [Eq. (8)], [2 + 3]-cycloaddition products are isolated [Eq. (11)] (20). With BEt$_3$

$$2\ Ar_2BN_3 \xrightarrow{-N_2} \begin{array}{c} Ar_2B\diagdown_{N}\diagdown^{N}\diagdown_{N} \\ \diagdown_{B}\diagup^{N} \\ ArAr \end{array} \qquad (11)$$

Ar = Mes, C$_6$F$_5$

as the trapping agent, products in accord with Eq. (9) were found, pointing to iminoboranes ArBNAr as intermediates for all four aryl groups under investigation. But in the case of phenyl and pentafluorophenyl, a second product in accord with Eq. (10) is formed as the major product, indicating two different mechanisms. The second one seems to involve greater steric requirements in the primary step and is inhibited, therefore, for Ar$_2$BN$_3$ with the bulky groups o-tolyl and mesityl. A difficulty for this interpretation comes from the observation that a bulky agent like BsBu$_3$ does trap iminoboranes ArBNAr, but predominantly (Ar = Ph) or even exclusively (Ar = C$_6$F$_5$) in accord with Eq. (9).

[Silyl(silyloxy)amino]boranes decompose in the liquid phase to a mixture of products [Eq. (12)] (18, 19). Again, the formation of the

$$R_2B=N\underset{SiMe_3}{\overset{OSiMe_3}{\diagup}} \xrightarrow{-(Me_3Si)_2O} R_2B\dot{-}NR\dot{-}BR\dot{-}N\underset{SiMe_3}{\overset{OSiMe_3}{\diagup}} \quad \tfrac{1}{3}(RBNR)_3 \qquad (12)$$

R = Me, Et, Pr, iPr, iBu, PhCH$_2$

product with the BNBN chain was explained by the presence of intermediate iminoboranes, which had been trapped by the starting borane. Again, the same product was found by aminoboration of the isolated iminoboranes with the reactants of Eq. (12) (Section V,C). Once more, BEt_3 proved to be a trapping agent. For the role of PhN_3 and Me_3SiN_3 as reagents for trapping iminoboranes, we refer to the above-cited literature (see also Sections V,C and VI,B).

There is some evidence for cyclic iminoboranes as intermediates. When cyclic chloroboranes react with Me_3SiN_3 at room temperature (i.e., the normal route for attaching an N_3 group to boron), evolution of N_2 is observed and one of the products of Eqs. (13) and (14) is isolated,

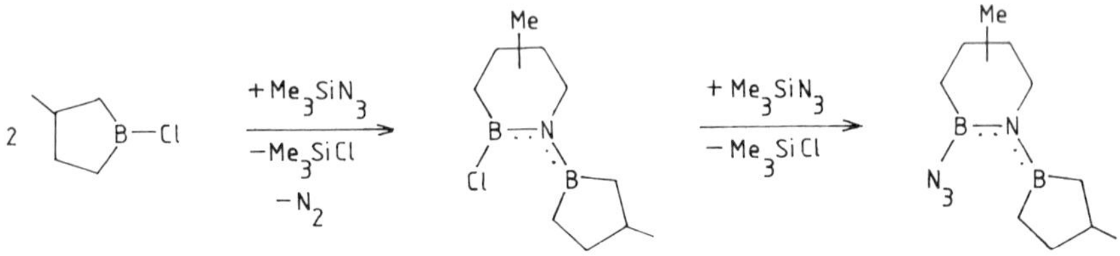

(13)

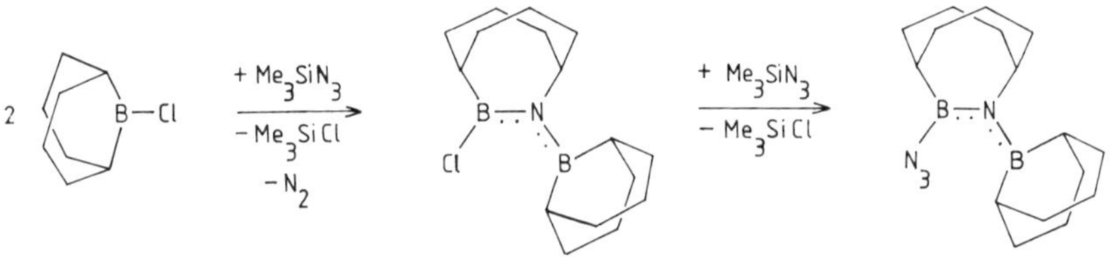

(14)

depending on the amount of Me_3SiN_3 (19). The most plausible mechanistic interpretation involves as an initial step the formation of the corresponding cyclic azidoboranes, which are unstable at room temperature. Elimination of N_2 is accompanied by ring expansion, yielding the cyclic iminoboranes **I** and **II**, respectively. By analogy with cycloalkynes having less than eight ring members, **I** and **II** are expected to be extremely reactive and to be chloroborated immediately by the starting borane.

The chloroborane reactants of Eqs. (13) and (14) can be aminated to isolable [silyl(silyloxy)amino]boranes, which decompose at 120 and 70°C, respectively, to give a mixture of products, three of which were identified in both cases [Eqs. (15) and (16)] (*19*). Again, the iminoboranes

$$(15)$$

$$(16)$$

I and **II** are postulated as reactive intermediates, adding, rather unspecifically, both the primary elimination product $(Me_3Si)_2O$ ("oxysilation") and the reactant ("aminoboration"). Performing the decomposition in an excess of siloxane $(Me_3Si)_2O$ gives only the oxysilation product.

Finally, three examples are reported in which iminoboranes as intermediates do not react with trapping agents but stabilize themselves intramolecularly in the gas phase during a hot tube procedure [Eqs. (17)–(19)]. A prerequisite is the steric availability of side groups with respect to the BN bonds (*9, 21*). The products are well established either by solvolytic degradation (*9*) or by X-ray analysis (*21*). Note that

$$(17)$$

$$\text{(18)}$$

$$\text{(19)}$$

the elimination of N_2 in Eqs. (18) and (19) is accompanied by the migration of an amino group, replacing a strong B—N by a weak N—N bond! Iminoboranes are the plausible intermediates in all three cases. The stabilization involves addition of a C—H bond to the iminoborane B—N bond ("alkylohydration"), provided that sterically fixed methyl groups are available [Eqs. (17), (18)]. In the course of Eq. (19), the iminoborane is presumed to eliminate propene via a six-membered transition state, leaving an azo compound that readily rearranges to the finally isolated hydrazone [Eq. (19a)].

$$\text{(19a)}$$

When thermolyzing azido compounds, the question arises as to whether an observed rearrangement is achieved in one step, concerted to the elimination of N_2, or whether there are two steps, elimination and rearrangement, with nitrenes as intermediates. Generally, thermolysis of azidoboranes is more probably a concerted process (22). The decomposition of $(iPr_2N)_2BN_3$, however, carried out at 450°C, gives a product, besides the main product of Eq. (19), that has to be interpreted in terms of a nitrene stabilized by an intramolecular CH insertion [Eq. (19b)].

$$\cdots \xrightarrow{-N_2} \left\{ \begin{array}{c} \text{[structure with B≡N and two N-iPr groups]} \end{array} \right\} \longrightarrow \text{[cyclic product structure]} \quad (19b)$$

III. The Structure of Iminoboranes

A. Theoretical Evidence

There are several nonempirical theoretical approaches to HBNH, starting from different bases (*23, 24*). The optimum arrangement of the atoms turns out to be a linear chain, point group $C_{\infty v}$, with BN bond lengths of 127 (*23*) and 123 pm (*24*), respectively. The highest occupied molecular orbitals (HOMOs) are the degenerate BN π-orbitals [orbital energies: -12.7 eV (*23*), -11.0 eV (*24*)], followed by the BN σ-orbital [-16.6 eV (*23*), -15.6 eV (*24*)]. The lowest unoccupied molecular orbitals (LUMOs) are the degenerate π^*-orbitals.

The BN bonding energy was found to be 88 kcal/mol. When compared with 94 kcal/mol for the isoelectronic ethyne, the B—N bond in HBNH seems to be somewhat weaker than a C≡C triple bond. (Both values were calculated without taking electron correlation into account and would be lower, therefore, than values from experiments; this does not limit a qualitative comparison.) The total BN overlap population was found to be 0.765, while the corresponding values for a B—N bond in the cyclooligomers (HBNH)$_2$ and (HBNH)$_3$ are 0.392 and 0.425, respectively.

Apparently there is a substantial contribution to the B—N bond of HBNH from one σ- and two orthogonal π-bonds. Expressed in simple terms, there is a B≡N triple bond in iminoboranes. Concerning a structural formula for HBNH, the real situation is represented best by H—B≡N—H.

The compound HBNH is expected to be thermodynamically unstable toward oligomerization. The cyclodimer, (HBNH)$_2$, isoelectronic with cyclobutadiene, is found to be more stable by 63.3 kcal/mol, and the cyclotrimer, (HBNH)$_3$, by 193.8 kcal/mol than 2 or 3 mol of HBNH, respectively (*23*). The oligomerization energy is due to an increase in the number of σ-bonds at the expense of the relatively weak π-bonds. Note that the situation is the same for acetylene. Oligomers like cyclobutadiene, benzene, benzvalene, and cyclooctatetraene are thermodynamically more stable than acetylene, and the formation of polyacetylene is also an exothermic reaction. Certainly, benzene is the most stable product of the stabilization of acetylene. In the corresponding BN case, the calculated oligomerization energies prove the cyclotrimer also to be more stable than the cyclodimer. Ring strain effects are responsible for that, and, at least in carbon ring systems, the cyclic delocalization of π-electrons plays an important role, according to Hückel's rule.

From gross electronic populations, the excess atomic charges can be derived as follows:

H—	—B	≡≡≡N—	—H	
0.03	0.26	−0.62	0.32	(23)
−0.03	0.21	−0.43	0.25	(24)

Apart from differences in the calculated values, the B—N bond in HBNH is a polar bond with a $\delta+$ charge on boron and a $\delta-$ charge on nitrogen. All three bonds exhibit some ionic character. Thus each of the π-bonds was found to be 39% ionic. This polarity does not imply that the π-bonds are substantially weakened; the ionic character reduces the π-bond strength from 100 to 89%. Applying the concept of formal charges to HBNH, a formula results that does not only overemphasize the amount of polarity, but also gives the wrong sign: H—$\bar{B}$≡$\overset{+}{N}$—H. Such a formalism has no physical meaning and cannot be recommended, as the only advantage seems to be increased clarity in electron counting. The dipole moment of HBNH was estimated to be 0.86 D (24).

A striking difference between alkynes and iminoboranes appears to be their kinetic stability. As was pointed out in Section II, iminoboranes are metastable, in general, at temperatures far below room temperature. Alkynes are also metastable, but their stabilization requires either high temperature or effective catalysts. We assume the polarity of the B—N bond to be a chief reason for these differences. This idea is supported by the observation that strongly polar alkynes (e.g., FC≡CH, FC≡CtBu) do oligomerize or polymerize at room temperature quite rapidly (25). Polar additions will generally be the predominant reaction for iminoboranes (Sections V,VI).

Considering derivatives of the "parent compound," HBNH, a MNDO/1 calculation was done on the dimethyl derivative, MeBNMe (11). Again, a linear CBNC chain turned out to represent the optimal geometry, point group C_{3v}, with a B—N bond distance of 119 pm. The HOMOs, once more, are two degenerate BN π-orbitals (-11.3 eV), followed by degenerate CH$_3$ σ-orbitals (-14.2 eV) and by the BN σ-orbital (-14.3 eV). A small dipole moment, 0.14 D, was proposed.

B. X-Ray Structural Evidence

The crystal and molecular structure of a single crystal of tBuBNtBu was investigated by X-ray methods at $-85°C$ (*11*). The substance crystallizes in the space group *Pnma*, isostructural with the isoelectronic alkyne tBuCCtBu (*26*). Boron and nitrogen atoms are disordered. The linear CBNC skeleton is framed by six methyl groups in an eclipsed conformational arrangement, leading to the point group C_{3v} for the iminoborane and D_{3h} for the alkyne. The central bond lengths were found to be 125.8 pm (BN) and 118.0 pm (CC).

In a recent summary on C≡C bond lengths, widespread values for 23 different bonds were cited with an average value of 118.5 pm, referring to X-ray determinations at room temperature; bond lengths from electron diffraction data or from microwave analysis were found to be distinctly longer (*27*). For a mostly qualitative general discussion we take 118 pm as typical for a C≡C bond. Apart from exotic examples, bond lengths for C—C single and C=C double bonds in crystals are found in a quite narrow range; values of 154 pm and 133 pm, respectively, can generally be accepted.

In noncyclic crystalline amine–boranes with noncyclic components BX_3 and NR_3, the upper limit of known B—N bond lengths is 160.2 pm in Br_3BNMe_3 (*28*), and the smallest value is 157.5 in Cl_3BNMe_3 (*29*); an average B—N bond length of 158.7 pm results, including seven further amine–boranes and including BX_3 components like BH_3, BF_3, $BH_2(NCS)$, and $[BHBrPic]^+$ (Pic = 4-picoline). We have to mention that in cyclic amine–boranes as well as in amine–boranes with cyclic components, larger BN distances are reported (e.g., 166.2 pm in the adduct from borane, BH_3, and urotropine (*30*)]. Boron–nitrogen bonds in gaseous amine–boranes seem to be larger, too; the two values 158.5 and 163.6 pm are found for the same substance, F_3B—NMe_3, in its crystalline and in its gaseous state, respectively (*31, 32*). For purposes of a gross comparison we take 159 pm as typical for the B—N single bond.

In order to obtain a typical bond length for the B=N double bond, we take into account aminoboranes X_2BNR_2 with planar tricoordinated boron and nitrogen atoms. With this restriction, diaminoboron cations with a dicoordinated boron atom [e.g., $[Me_2N=B=NC_9H_{18}]^+$ (*33*)] or (alkylidenamino)boranes with a dicoordinated nitrogen atom [e.g. $Mes_2B=N=CPh_2$ (Mes = mesityl) (*34*)], are not considered; these molecules may have particularly short B—N distances, the extreme value being 130 pm (*33*). Molecules where boron and nitrogen are in an exceptional steric situation are also not considered [e.g., the trapezoidal molecule **III**, whose B=N bond length is only 134 pm (*16*)]. On the other hand, extremely long B—N distances may be found when such groups are bonded to nitrogen that compete with the boron atom for the nitrogen π-electrons [e.g. $F_2B{=}N(SiH_3)_2$, $d(B{=}N) = 149.6$ pm (from electron diffraction) (35)]. Finally, an extra-long B—N bond might be found for the central bond in molecules with BNBN chains, which are isoelectronic with 1,3-dienes [e.g. **IV**, $d(BN)_\beta = 148.8$ pm (36)].

III   IV

The B—N distances for five noncyclic aminoboranes were reported with an average distance of 141.4 pm, the extreme values being 137.9 pm in $Cl_2B{=}NPh_2$ (37) and 143.9 pm in $B(NMe_2)_3$ (38). The relatively large B—N bond length for the triaminoborane can be well understood, because three nitrogen lone pairs have to share one vacant boron p-orbital. The crystal and molecular structures of 30 cyclic aminoboranes are reported to have B—N bond lengths similar to those for noncyclic aminoboranes, the average value being 141.8 pm. Borazines $(XBNR)_3$, which were included in the averaging procedure, tend to have longer B—N bonds (~143 pm) than the average value. The difference from the average value would be rather small if the C—C bond lengths in the isoelectronic species alkene and benzene, 133 and 140 pm, respectively, were considered. One might have in mind, however, that too detailed conclusions from bond lengths could become meaningless for methodical reasons, because of crystal effects in the case of X-ray analysis, because of the lack of simple correlations between bond lengths and electronic properties, etc. The highest values in averaging the B—N bond lengths of cyclic aminoboranes were contributed by three diazadiboretidines $(XBNR)_2$, the cyclodimers of iminoboranes (Section IV,B), with bond lengths close to 145 pm.

Though adoption of a "typical" B=N bond length may be questionable, we suggest 141 pm as a possible value, referring to noncyclic aminoboranes. The most questionable course is to generalize the one experimentally established B≡N triple bond distance of 126 pm. On the other hand, the steric situation at least will not be too different, if the substituents R and X in XB≡NR, distant from each other, are altered. Thus we take 126 pm as a typical B≡N triple bond distance. The preceding discussion is summarized in Table II.

TABLE II

COMPARISON OF CC AND BN BOND LENGTHS

CC species	d (pm)	Δd (pm)	BN species	d (pm)	Δd (pm)
—C—C—	154		—B—N—	158	
		21			17
C=C	133		B=N	141	
		15			15
—C≡C—	118		—B≡N—	126	

The difference between a single and a double bond distance is nearly 20 pm but it is only 15 pm between a double and a triple bond. Boron–nitrogen bonds are longer than the corresponding C—C bonds, the difference being nearly 5 pm for the single bonds and nearly 8 pm for the double and triple bonds.

C. SPECTROSCOPIC EVIDENCE

Much is known empirically on ^{11}B-NMR data (39). Restricting boron compounds to noncyclic molecules with no or one nitrogen atom and two, three, or four carbon atoms bonded to boron and omitting sterically overcrowded alkyl groups R and R′, we find four classes with the following typical chemical shift ranges ($Et_2O \cdot BF_3$ as the external standard):

$[R'_4B]^-$	$R'_3B—NR_3$	$R'_2B=NR_2$	R'_3B	
−22 to −16	−9 to +5	+42 to +49	+83 to +88	ppm

Trialkylboranes represent the smallest electronic shielding of the ^{11}B nucleus; only three σ-bonds are available for a sextet boron atom. One additional π-bond in aminoboranes increases the shielding effect. Four σ-bonds in amine–boranes exhibit still stronger shielding, and an additional negative charge in tetraalkyl borates brings a maximum shielding in this series. Iminoboranes R′B≡NR have two σ- and two π-bonds. The typical ^{11}B-NMR shifts in the range 2.3–6.3 ppm (Section II,A) apparently cannot be explained in the above oversimplified σ/π bond picture. The orthogonal π-electron distribution may cause an unusual effect of diamagnetic anisotropy. A theoretical elucidation would be desirable.

A vibrational analysis is reported for HBNH (8) and tBuBNtBu (40). Infrared data for the parent compound in an argon matrix were obtained with 10 isotopically labeled species, including the isotopes ^{1}H, D, ^{10}B, ^{11}B, ^{14}N, and ^{15}N. With restriction to the species H^{11}B^{14}NH, most common from natural material, two Σ^+ stretching modes were found at 3700 and 1785 cm^{-1}, which were readily assigned to the asymmetric hydrogen stretch v_1 (with more NH character) and to the BN stretching vibration v_3. The third Σ^+ mode (i.e., the symmetric hydrogen stretch v_2, with more BH character), could not be detected. All three Σ^+ vibrations were calculated from a set of three force constants; of the resulting wave numbers, 3700, 2800, and 1785 cm^{-1}, two are in ideal accord with the observed values. The relatively weak intensity of v_3 and the vanishing intensity of v_2 lead to a very small dipole moment for HBNH, so that the vibrations v_2 and v_3 become comparable to the IR-inactive Σ_g^+ vibrations of the isoelectronic molecule HCCH. One of the two expected Π bending frequencies was detected at 460 cm^{-1}, representing more BNH than HBN bending character. The best force constant k(BN) turned out to be 13.14 N/m.

The IR and Raman spectra of liquid tBuBNtBu with fairly pure ^{10}B and in natural isotopic abundance were recorded, together with the corresponding spectra of the isoelectronic alkyne tBuCCtBu. In accord with a normal coordinate analysis, the fundamental frequencies were assigned in terms of a linear central skeleton and a staggered arrangement of the methyl groups. The resulting D_{3d} symmetry of the alkyne was found to agree with the exclusion principle for IR and Raman intensities in the presence of a center of symmetry. Though there is no such center in the iminoborane with C_{3v} symmetry, the same exclusion principle is at least indicated by corresponding strong or weak intensities and vice versa in both spectra, representing evidence, once more, for a small dipole moment of the iminoborane and a distinct structural similarity of the two isoelectronic species. The more unfavorable eclipsed conformation of both molecules in the solid state seems to be evoked by crystal lattice forces. The ^{11}B—N stretching frequency was found at 2009 cm^{-1}; the C≡C Raman band of the corresponding alkyne appears at 2226 cm^{-1}. The large difference of 224 cm^{-1} in the ^{11}BN frequencies of HBNH and tBuBNtBu is due to the strong coupling between the symmetric stretching vibrations v_2 and v_3 of HBNH and does not imply a difference in the bond strengths. A force constant k(BN) = 12.79 N/m was the best fitting value for tBuBNtBu.

The BN force constants of HBNH and tBuBNtBu (13.14 and 12.79 N/m, respectively) are similar in magnitude. The corresponding CC force constants of HCCH and tBuCCtBu [15.59 (41) and 15.58 N/m, respectively] indicate the C≡C triple bond to be stronger than the B≡N triple bond. From average values [k(B≡N) = 12.9 and k(C≡C) = 15.6 N/m] we find k(B≡N) = 0.83k(C≡C). The same trend was deduced from MO calculations and from bond lengths (Sections III,A,B). The relationship between alkynes and iminoboranes can be compared to the relationship between the isoelectronic molecules N_2 and CO: The N≡N bond (109.76 pm) is shorter than the C≡O bond (112.82 pm) (42), and the force constant of N_2 (22.98 N/cm) is larger than that of CO (18.47 N/cm) (43), according to k(CO) = 0.80k(N_2). This trend does not hold for the dissociation energy, which is larger for CO than for N_2.

The photoelectron spectrum of tBuBNtBu shows the first ionization energy to be 9.35 eV (11), compared to 9.05 eV for the corresponding alkyne tBuCCtBu (44). The photoelectron is expected to be one of the four π-electrons.

D. Dipole Moments

A value of 0.20 D was found for the dipole moment of tBuBNtBu in cyclohexane (11). Theoretically estimated values are 0.86 D for HBNH and 0.14 D for MeBNMe with $\delta+$ on boron and $\delta-$ on nitrogen (Section III,A). The vibrational analysis of tBuBNtBu indicates a small dipole moment (Section III,C). If conclusions from reaction paths to ground states are permitted, the dipole direction in all known iminoboranes will be the one predicted by theory, since addition of polar agents AY to the BN bond principally directs the positively charged fragment A to nitrogen and the negatively charged fragment Y to boron (Sections IV–VI).

More generally, "symmetric" iminoboranes RBNR, with identical ligands at both triply bonded atoms, seem to have a small dipole moment in common with carbon monoxide, CO, whose dipole moment is 0.112 D. In spite of being small, the polarity of the BN bond as well as that of the CO bond is large enough to make iminoboranes as well as carbon monoxide more reactive toward polar addition than the isoelectronic alkynes and dinitrogen, respectively. A difference between RBNR and CO is the direction of the dipole; in CO, the more electronegative oxygen bears the small positive charge (45), which happens to be in accord with the sign of the formal charges in the structural formula [$\overset{-}{C}$≡$\overset{+}{O}$]. Although not theoretically sound, an experimentally verifiable conclusion would be that the carbon atom in CO works as the Lewis base center, which is well established for BH_3 or certain transition metal compounds as the corresponding Lewis acids.

E. Conclusions Concerning Structural Formulas

It can be concluded from the discussions in Sections IIIA–D that $B=N$ double bonds in aminoboranes and $B\equiv N$ triple bonds in iminoboranes represent a realistic picture. It is here recommended, therefore, to indicate these bonds in structural formulas as usual, but to omit erroneous formal charges, e.g., amine–borane: X_3B-NR_3; aminoborane: $X_2B=NR_2$; iminoborane: $XB\equiv NR$. [Note that $R_3N \cdot BX_3$ is recommended as the correct molecular formula for amine–boranes (46), but one is not bound to rules in constructing structural formulas, e.g., X_3B-NR_3.]

Arrows instead of dashes for the representation of so-called coordinative covalent bonds call for mention. The history of bond formation may not be represented in a formula. Furthermore, in formulas like $[R_3N \rightarrow BH_2-NR_3]^+$ or $XB\equiv NR$, for example, two symmetrically equivalent bonds are represented by a plotting procedure different for each of the two, a rather illogical way, particularly, since it is uncertain which of the two bonds, if either, had been the coordinative one during the formation.

There is no general objection to writing down mesomeric structures, e.g.,

$$\{X_2B-\bar{N}R_2 \longleftrightarrow X_2B=NR_2\}$$

but the less complicated method of omitting the structure of obviously little weight seems to be preferable. When a π-electron pair is delocalized over more than two atoms, it is preferable to draw one formula with dotted lines along the area of delocalization. For example, in the case of diaminoboranes $XB(NR_2)_2$:

$$R_2N\cdots B(X)\cdots NR_2,$$

is preferred over a pair of mesomeric structures:

$$\{R_2N=B(X)-NR_2 \longleftrightarrow R_2N-R(X)=NR_2\}$$

The cyclic delocalization of π-electrons in diazadiboretidines $(XBNR)_2$, or borazines, $(XBNR)_3$, etc., will be indicated by dotted lines over bonds, again favored over mesomeric structures.

IV. Oligomerization of Iminoboranes

A. Survey

The activation barrier for the oligomerization of alkynes may be overcome thermally or catalytically. Because the classical thermal transformation of ethyne into benzene in a hot tube is rather ineffective (47), more emphasis has been placed on working out catalytic routes for the synthesis of linear oligomers, cyclooligomers, and polymers. Transition metal compounds have proved to act as effective catalysts in homogeneous as well as in heterogeneous processes (48).

The stabilization of iminoboranes can yield five different types of products: cyclodimers (1,3,2,4-diazadiboretidines, **Di**), cyclotrimers (borazines, **Tr**), bicyclotrimers (Dewar borazines, **Tr'**), cyclotetramers (octahydro-1,3,5,7-tetraza-2,4,6,8-tetraborocines, **Te**), and polymers (polyiminoboranes, **Po**); these substances are isoelectronic with cyclobutadienes, benzenes, Dewar benzenes, cyclooctatetraenes, and polyalkynes, respectively, which are all known to be products of the thermodynamic stabilization of alkynes.

A correlation between the iminoboranes and their thermal stabilization products is given in Table III. *Thermal,* in this context, means "at room temperature." Mixtures of two products can be separated either by extraction of the soluble component (**Tr/Po**) or by distillation

TABLE III

Products from the Stabilization of Iminoboranes XBNR

X	R	Product of stabilization — Thermal	Product of stabilization — Catalytic	Reference
Me	Me	Tr	Tr	19
Me	tBu	Tr	Di ⇌ Te	13
Et	Et	Tr, Po		18, 19
Et	tBu	Tr	Di	10, 13
Pr	tBu	Tr	Di	10, 13
iPr	iPr	Tr	Di ⇌ Te	17, 19
iPr	tBu	Tr'	Di	12
Bu	tBu	Tr	Di	10, 49
iBu	iBu	Tr, Po		17
sBu	sBu	Di, Tr	Di	17, 19
sBu	tBu	Di, Tr'		19
tBu	tBu	Di		11
F_5C_6	tBu	Di		9
(2,2,6,6-tetramethylpiperidin-1-yl)	tBu	Di		16
tBu(Me$_3$Si)N–	tBu	Di		14

(**Di/Tr, Di/Tr'**). A general result is that sterically normal ligands X and R cause the trimerization of iminoboranes to borazines, whereas steric strain by X and R can make the cyclodimers more favorable. Obviously, the space available for the ligands of planar rings decreases with the number of ring members. A borderline situation arises with two of the doubly α-branched sec-butyl groups as ligands; in this case a mixture of the cyclodimer and -trimer is formed.

One doubly α-branched and one triply α-branched ligand mark the very special situation where Dewar borazines become stable (iPr/tBu, sBu/tBu); these ligands are small enough to permit more than a cyclodimerization but are too big to make borazines favorable. Going one step further to two triply α-branched ligands (tBu), the cyclodimer becomes the only possible stabilization product. Polymers (RBNR)$_n$ together with borazines are the products of heating metastable iminoboranes with two α-unbranched groups R. A corresponding mixture was recovered from the hot tube thermolysis of R_2BN_3 (17) and of $R_2BN(SiMe_3)OSiMe_3$ (*18*) (R = Pr, Bu), without isolating and characterizing the monomers. Little is known about the polymers. They are colorless, waxlike materials, which are insoluble in all kinds of organic solvents and can be stored in the open air for some time. The mass spectrometric fragmentation gives cations $(RBNR)_m^+$ with a maximum value of $m = 5$; nothing is known, however, about the true size of what are called *polymers*. Their insolubility and lack of swelling capability (*19*) make a truly polymeric structure not unlikely. Polymers could not be detected during the stabilization of MeBNMe (*19*).

Catalysts, working below 0°C, have been found that induce the formation of cyclodimers from those iminoboranes which give borazines at room temperature. Cymantrene, $CpMn(CO)_3$, and similar coordination compounds of transition metals, exhibit catalytic activity (*19*), but the most effective catalyst turned out to be *tert*-butylisonitrile, *t*BuNC (*12, 13, 49*). Four cyclodimers, which were available only by such a catalytic procedure (Table III), proved to be thermally stable toward transformation into the corresponding borazine, which obviously will be thermodynamically more stable. In two cases an equilibrium mixture of the cyclodimer and the cyclotetramer was obtained (Table III); neither the cyclodimer nor the cyclotetramer could be transformed into the corresponding borazine. Low-temperature NMR data indicate a possible mechanism for the catalytic activity of *t*BuNC according to Eq. (20) (*19*).

$$2\ R'BNR \xrightarrow{+\ tBuNC} \left\{ \begin{array}{c} R\ \ \ R' \\ N\!\!=\!\!B \\ R'B\diagdown\!\!NR \\ C \\ \parallel \\ N\diagdown_{tBu} \end{array} \right\} \xrightarrow{-\ tBuNC} \begin{array}{c} R' \\ B \\ RN\diagdown\!\!NR \\ B \\ R' \end{array} \quad (20)$$

B. The Cyclodimers

Diazadiboretidines are isoelectronic with cyclobutadienes. A rectangular D_{2h} structure with localized π-bonds is indicated for cyclobutadiene by theory and is strongly supported by experimental evidence (*50*). Being a highly reactive diene as well as a strong dienophile, cyclobutadiene will readily undergo a Diels–Alder cycloaddition with itself and is, therefore, unstable. The same is true for its derivatives, unless the ring ligands exhibit particular steric or electronic effects.

For example, the sterically overcrowded tetra-*tert*-butylcyclobutadiene, whose photochemical transformation into the corresponding tetrahedrane is of principal interest (51), is a rather stable substance with a nonplanar ring skeleton (52). Four ring ligands with clockwise opposed electronic effects define so called push–pull cyclobutadienes (e.g., **V**), which are storable at room temperature (53). The ring skeleton of **V** is rhombic, instead of rectangular, with an acute angle of 87.2° at the carbon atoms that bear the EtOOC groups (54). The π-electrons are delocalized over the ring and the four adjacent external bonds.

$$\text{EtOOC}-\underset{\underset{\underset{\text{NEt}_2}{|}}{C}}{C}\overset{\overset{\overset{\text{NEt}_2}{|}}{C}}{\diamondsuit}C-\text{COOEt}$$

V

In the homologous boron–nitrogen four-membered ring systems the ring atoms themselves, not the ligands, push and pull the electrons. The structures of four examples were analyzed in the crystalline state. Like its carbon homologue, the tetra-*tert*-butyl derivative shows slight deviations from a planar structure (11), but diazadiboretidines with sterically less outstanding ligands have a planar, rhombic ring skeleton with the acute angle at the nitrogen atoms, comparable to the push–pull cyclobutadiene (Table IV).

Ab initio calculations were reported for the parent compound, (HBNH)$_2$ (23, 56). A rhombic C$_{2v}$ structure with a B—N bond length of 147 pm and a B—N—B angle of 87° was predicted. The π-bonding

TABLE IV

STRUCTURE OF IMINOBORANE CYCLODIMERS (XBNR)$_2$

X	R	Mean ring bond lengths (pm)	Mean ring angles (degree)		Reference
			BNB	NBN	
Bu	*t*Bu	145.8	85.3	94.7	14
*t*Bu	*t*Bu	148.6	86.6	90.6	11
F$_5$C$_6$	*t*Bu	143.1	84.3	95.7	9
(Me$_3$Si)$_2$N	SiMe$_3$	145.4	82.2	97.8	55

energy is greater by 11 kcal/mol than for two isolated BN π-bonds, in contrast to (HCCH)$_2$, for which the delocalization energy is predicted to be negative (*56*). The ionization potential was calculated to be 10.8 (*23*) and 8.1 eV (*56*). A rhombic D$_{2h}$ structure with a B—N bond length of 146 pm, a B—N—B angle of 88°, and an ionization potential of 9.13 eV was calculated for (MeBNMe)$_2$ (*11*). In all calculations, the HOMO is predicted to be the π-orbital of b$_g$ symmetry, whose electron density is localized at the two nitrogen atoms. The photoelectron spectrum of (*t*BuBN*t*Bu)$_2$ shows the ionization energy to be 7.35 eV (*11*), not very different from the 6.35 eV reported for the corresponding (*t*BuCC*t*Bu)$_2$ (*57*). The small value, compared to the one calculated for (MeBNMe)$_2$, was ascribed to the inductive effect of the *tert*-butyl groups.

One could expect diazadiboretidines to be converted into Hückel aromatic systems either by adding or by subtracting one pair of π-electrons. The addition of two electrons to diazadiboretidines of the type (RBN*t*Bu)$_2$ can be achieved by the action of alkali metals. The dianions [(RBN*t*Bu)$_2$]$^{2-}$ are stable in solution and can be reconverted into the diazadiboretidines by oxidants. Because they contain six π-electrons, "aromatic character" may be attributed to the dianions (*19*). Cyclodimers of the type (R'BNR)$_2$ are also readily oxidized, but the adoption of an "aromatic" dication [(R'BNR)$_2$]$^{2+}$ as a product would be mere speculation at present.

Whereas push–pull cyclobutadienes are not formed by thermal cyclodimerization of the corresponding alkynes, there are six iminoboranes mentioned in Table III that undergo thermal cyclodimerization, and four further iminoboranes are likely to be formed as gas-phase intermediates before dimerizing to the well-characterized compounds [MesBN(SiMe$_3$)]$_2$ (*9*), [(F$_5$C$_6$)BN(SiMe$_3$)]$_2$ (*9*), [Me$_3$Si(*t*Bu)NBN*i*Pr]$_2$ (*14*), and [Me$_3$Si(*t*Bu)NBNBu]$_2$ (*14*).

For reason of symmetry conservation, a thermal concerted [2 + 2]-cycloaddition is forbidden as far as D$_{2h}$ symmetry can be assumed for the activated complex. This is not necessarily the case for the concerted dimerization of iminoboranes, in which the symmetry requirements seem to be essentially lowered. Nevertheless, a two-step mechanism according to Eq. (21) must be taken into account. The assumed intermediate in Eq. (21) contains a sextet boron atom with a linear coordination, which seems to be rather unfavorable; the aminoboron cation $[C_9H_{18}N=B-Me]^+$, containing a boron atom of that type, is reported, however, to be metastable at low temperatures (*33*).

C. THE CYCLOTRIMERS

Borazines, the normal products of the thermal stabilization of iminoboranes, constitute a well-characterized class of molecules (*58*). The parent compound, $(HBNH)_3$, has been known for 60 years (*59*). Some of the physical properties of borazine and benzene are so similar that borazine was called *inorganic benzene* (*60*). Many of the theoretical contributions concerned the degree of aromaticity of borazines. On the other hand, the first Dewar borazine, formed by trimerization of the iminoborane *i*PrBN*t*Bu, was reported in 1984 (*12*). In the crystalline state, the bicyclic skeleton is built from two trapezoids, joining the longer edge and including an angle of 115.2°. The common central edge forms an extra-long B—N single bond of 175.2 pm, whereas the two opposite short edges of the trapezoids indicate rather short B=N double bonds with a BN distance of 136.4 and 138.4 pm, respectively. A fluxional rearrangement of $(i\text{PrBN}t\text{Bu})_3$ in a solution of $CDCl_3$ was suggested [Eq. (22)], which was fast at 20°C on the NMR time scale. Though solutions of this rather insoluble Dewar borazine at low temperature were not available, the NMR spectra at 20°C differed from the expected spectra of the corresponding borazine, not by the number of signals, but by the chemical shifts, which were in accord with the shifts expected by averaging plausible Dewar borazine shifts.

$$\text{(22)}$$

A closely related Dewar borazine, $(t\text{BuBN}i\text{Pr})_3$, was prepared from the corresponding fluoroborazine $(\text{FBN}i\text{Pr})_3$ by substitution of all three fluorine atoms. This Dewar borazine is soluble at $-50°C$ in a mixture of $CDCl_3$ and CH_2Cl_2. The NMR spectra at that temperature correspond to the expected Dewar borazine structure, the signals coalescing at higher temperature (*61*). The activation energy for the valence isomerization [Eq. (22)] seems to be small. The transition state will have a structure not very different from a borazine structure. If only one methyl group in each *tert*-butyl group is replaced by a hydrogen atom, the Dewar borazine structure will become unfavorable, since the borazine (iPrBNiPr)$_3$ is a well-established stable substance, formed for instance by the thermal stabilization of monomeric iPrBNiPr (Table III). The compound (sBuBNtBu)$_3$ was also found to be a Dewar borazine from spectroscopic data, which were similar to those of (iPrBNtBu)$_3$ (*19*).

Dewar benzene derivatives have been known since 1962 (*62*), the parent compound, C_6H_6, since 1963 (*63*). The rearrangement of a normal Dewar benzene to the corresponding benzene is an exothermic reaction, but derivatives like hexamethyl Dewar benzene are metastable at room temperature (*64*), and the parent compound can be stored in a pyridine solution below 0°C. Strong steric strain can make the Dewar benzene more favorable than the benzene even thermodynamically; four *tert*-butyl groups together with two methoxycarbonyl groups can exhibit such a strain (*65*). Apparently, a similar situation is met with the Dewar borazines. A fluxional behavior, however, is not reported for the strained Dewar benzenes, a conversion by simple thermal opening of the central C–C bond being forbidden by reason of orbital symmetries. The structures of Dewar benzene and Dewar borazine are comparable. An electron diffraction study of hexamethyl Dewar benzene showed that the two equal trapezoid moieties include an angle of 124°; the lengths of the central CC bond and of the opposite C=C double bonds are 163 and 135 pm, respectively (*66*). With respect to iminoboranes, it is remarkable that the Dewar benzene C_6F_3tBu$_3$ is one of the products of the spontaneous oligomerization of the polar alkyne FC≡CtBu (*25*). A yield of 60–70% is reported for the synthesis of hexamethyl Dewar benzene from MeC≡CMe in the presence of AlCl$_3$ (*64*).

D. THE CYCLOTETRAMERS

Cyclooctatetraene, C_8H_8, has a D_{2d} tublike structure with four rather isolated double bonds. Cyclooctatetraenes can be the products of the catalytic cyclotetramerization of alkynes, and cyclobutadienes may be the intermediates. The BN homologues of cyclooctatetraenes have been known since 1962 (*67*). Like cyclooctatetraene, molecules of [(SCN)BNtBu]$_4$ were shown to have a tublike ring structure of S_4 symmetry with alternating bond lengths of 140 and 146 pm, the shorter ones perpendicular to the direction of the S_4 axis (*68*).

The equilibrium 2**Di** ⇌ **Te** was observed after a catalytic iminoborane stabilization in the case of two particular combinations of the ligands X and R: Me/tBu (*13*) and iPr/iPr (*19*) (Table III). The equilibrium strongly depends on temperature: At 20°C, only the cyclotetramers are detectable by NMR; at 70°C (Me/tBu) and 100°C (iPr/iPr), respectively, the cyclotetramers are completely transformed into the cyclodimers, and mixtures are found at intermediate temperatures. The fragmentation **Te** → 2**Di** can be achieved within a few minutes (Me/tBu) or more slowly (iPr/iPr), but the reverse reaction 2**Di** → **Te** at 20°C takes hours and needs to be completed by UV irradiation in the case of X/R = iPr/iPr. We develop a mechanistic proposal in Eq. (23): a Diels–Alder homologous cycloaddition, followed by rapid opening of two bonds between tetra-coordinated boron and nitrogen atoms.

$$\text{(23)}$$

Going from the ligand combination X/R = Me/tBu to the slightly larger set of ligands Et/tBu, the cyclodimer remains thermally stable; no cyclotetramer is observable. The transformation **Di** ⇌ **Te** seems to be very sensitive to the steric situation in the ligand sphere of the cyclodimers. On the other hand, diazadiboretidines with a set of smaller ligands (e.g., Me/Me, Et/Et) have never been isolated. They may exist as intermediates, but there will be a favorable route to the formation of borazines, possibly through the intermediates of Eq. (23), as indicated in Eq. (24).

$$\text{(24)}$$

E. INTERCONVERSIONS BETWEEN CYCLOOLIGOMERS

Interconversion of cyclodimers and cyclotetramers was dealt with in Section IV,D. A very special conversion, the photochemical isomerization of tetra-*tert*-butyldiazadiboretidine to the corresponding tetrahedrane, might be expected by analogy with the behavior of tetra-*tert*-butylcyclobutadiene (*51*); all attempts in this field, however, have failed (19). Conversion of the cyclodimer into the cyclotrimer may be possible in the case of unstable cyclodimers, as was pointed out in the preceding section, but once formed as storable products, such a conversion was not observed. The opposite is not true. The Dewar borazine $(t\text{BuBN}i\text{Pr})_3$ (Section IV,C) can be converted into the corresponding cyclodimer $(t\text{BuBN}i\text{Pr})_2$ by a thermal process at 200°C (61). The mechanism of such a process, $2\mathbf{Tr'} \to 3\mathbf{Di}$, is unknown. Thermodynamically, the bicyclotrimer seems to be only slightly better in energy, so that an entropy term may provide the driving force at 200°C.

Apart from a real interconversion, cyclodimers can be transformed into borazines, however, by addition of iminoboranes [Eq. (25)]. In order to carry out such a reaction, a solution of the iminoborane in a dropping funnel, kept at $-80°C$, is slowly dropped into a solution of the cyclodimer at 50°C. The yield of borazines is quantitative. The procedure can be applied to components with the same set of ligands, but different sets may also be applied, permitting the synthesis of borazines with an unsymmetric arrangement of more than two different ligands (13, 19).

$$\text{[cyclodimer]} + \text{X'B}\equiv\text{NR'} \longrightarrow \text{[borazine]} \quad (25)$$

X:	Et	Pr	Bu	sBu	Et	Pr	iPr	Bu	iBu	sBu
R:	tBu	tBu	tBu	sBu	tBu	tBu	tBu	tBu	tBu	tBu
X':	Et	Pr	Bu	iPr	iPr	iPr	iPr	iPr	iPr	iPr
R':	tBu	tBu	tBu	iPr	iPr	iPr	iPr	iPr	iPr	iPr

By analogy with Eq. (25), the cyclodimer $(i\text{PrBN}t\text{Bu})_2$ will give the corresponding Dewar borazine, if reacted with $i\text{PrBN}t\text{Bu}$ (12). If the iminoborane $i\text{PrBN}i\text{Pr}$, instead of $i\text{PrBN}t\text{Bu}$, is added to the same cyclodimer, however, only the normal borazine is formed (19). Whether the normal or the Dewar borazine will be more stable depends on the difference of one single methyl group. The cyclodimer $(s\text{BuBN}t\text{Bu})_2$ behaves in the same way as $(i\text{PrBN}t\text{Bu})_2$: Only the normal borazine is formed by addition of $i\text{PrBN}i\text{Pr}$. Equation (25) can be interpreted as a [4 + 2]-cycloaddition giving the corresponding Dewar borazine as an intermediate, which will be readily transformed into the borazine if ligands with normal steric requirements are present.

Equation (25) may shed light on the general path of the iminoborane oligomerization. I propose the formation of cyclodimers to be the first stage of such oligomerizations. If a cyclodimer is stable to an excess of iminoborane, it will be isolated (Table III). Otherwise the cyclodimer is attacked by the excess iminoborane according to Eq. (25), and the borazine is formed via the Dewar borazine. In special cases, the Dewar borazine will be the final product. The first step determines the rate of such a sequence of reactions. If the cyclodimerization step becomes relatively fast, so that the first and the second step are comparable in rate, both the cyclodimer and the cyclotrimer will be found; this is true for the thermal stabilization of sBuBNsBu. Catalysts for the cyclodimerization make the first step more rapid than the second one.

A plausible alternative mechanism involves as a first step the formation of a linear dimer, XB=NR—BX=NR, according to the left part of Eq. (21). This linear dimer will be more stable in entropy but less stable in energy than the corresponding cyclodimer. The second step would be addition of iminoborane to the open-chain dimer giving the borazine in either a concerted or a two-step mechanism. The intramolecular cyclization of the open-chain dimer, according to the right-hand side of Eq. (21), would compete addition of an iminoborane. We cannot definitely exclude a mechanism via open-chain dimers.

Polymers are formed, together with borazines, from iminoboranes RBNR with α-unbranched alkyl groups. By absolute control of the temperature, the stabilization would presumably be directed toward the borazine. Loss of thermal control will cause a loss of kinetic control, so that hot iminoborane molecules will trimerize or polymerize rather unspecifically.

There is one further remarkable interconversion. The four polymers $(RBNR)_n$ (mentioned in Section IV,A) are thermally stable, except for $(EtBNEt)_n$, which can be transformed into the borazine $(EtBNEt)_3$ at 150°C (*18*). Such a depolymerization will proceed without considerable change of energy but with a substantial gain in entropy. For kinetic reasons, it cannot proceed with alkyl groups larger than ethyl.

V. Polar Additions to Iminoboranes

A. Addition of Lewis Acids and Bases

Neutral Lewis acids can be bonded to the nitrogen atom of the aminoiminoborane $C_9H_{18}NBN t Bu$ (*69*) [Eq. (26)]. The product is related to the well-known diaminoboron cations of the type $[C_9H_{18}N=B=$

$NR_2]^+$ (33); the positive charge is compensated intramolecularly, however, and not by a separate anion.

$$\text{pip-N}\overset{+}{\text{---}}B{\equiv}N\textit{t}Bu + A \longrightarrow \text{pip-}\overset{\oplus}{N}{=}B{=}\overset{\ominus}{N}\begin{matrix}A^{\ominus}\\ \textit{t}Bu\end{matrix} \quad (26)$$

$A = AlCl_3, AlBr_3, GaCl_3$

Efforts to add Lewis acids to dialkyliminoboranes R'BNR were not so successful, as would be expected, since betaine structures of the type R'B$\overset{+}{=}$NR$\overset{-}{-}$A with an unfavorable linear sextet boron atom would be formed (19, 33). Equation (26) is restricted to iminoboranes XBNR with a π-electron donating group X.

Polar reagents AY (Section V,B-D) generally attack both triply bonded atoms of iminoboranes to yield aminoboranes [Eq. (27a)]. In special cases, the cationic fragment A^+ of AY is added to the nitrogen atom and Y^- remains a separate anion [Eq. (27b)]; such a reaction path seems to be governed by steric factors, but seems also to be restricted to aminoiminoboranes (70).

$$\text{(27)}$$

(b): A—Y = Me_3Si—I, Me_3Si—$OS(CF_3)O_2$

The typical ^{11}B-NMR signals of iminoboranes are essentially unaltered when iminoboranes are dissolved in liquids with Lewis base activity (e.g., tertiary amines, tetrahydrofuran) (19). I conclude that equilibria like Eq. (28) are shifted far to the left, even in the presence of an excess of the Lewis base D.

$$XB{\equiv}NR + D \rightleftharpoons \begin{matrix}{}^{\oplus}D\\ \overset{\ominus}{B}{=}N\\ X \quad R\end{matrix} \quad (28)$$

B. Addition of Protic Agents

Each of six protic agents was added to both of two representative iminoboranes, iPrB≡NiPr and BuB≡NtBu; the expected 12 aminoboranes were isolated, chiefly in good yield [Eq. (29)] (71). The yield of distilled pure products may be smaller, but primarily the addition of protic agents is a quantitative reaction, fast even far below 0°C. This means a distinct difference to the slow addition of the same protic agents to alkynes which affords catalytic support at temperatures above 0°C.

$$R'B \equiv NR \ + \ H-Y \ \longrightarrow \ \underset{R'}{\overset{Y}{\diagdown}} B = N \underset{R}{\overset{H}{\diagup}} \qquad (29)$$

HY = HCl, tBuOH, Et$_2$NH, iPr$_2$NH, tBuNH$_2$, (Me$_3$Si)$_2$NH

Analogous products were recovered from the addition of HCl, iPrOH, and tBuNH$_2$ to the aminoiminoborane Me$_3$Si-(tBu)N⋯B≡NtBu (14) and from the addition of three acids (CH$_3$COOH, CF$_3$COOH, CF$_3$SO$_3$H), five alcohols ROH (R = Me, iPr, tBu, Ph, 2,4,6-tBu$_3$C$_6$H$_2$), four amines RNH$_2$ (R = H, iPr, tBu, Ph), five secondary amines (Me$_2$NH, pyrrole, pyrrolidine, pyrazole, imidazole), and two hydrazines (Me$_2$NNH$_2$, MeHNNHMe) to the aminoiminoborane C$_9$H$_{18}$N⋯B≡NtBu (70, 72).

Apparently, Eq. (29) represents a polar nonradical addition. If a two-step mechanism is conceived, intermediates of the type [XB=NRH]$^+$ will be reasonable, though such cations proved to be rather unstable as isolated species (unless X represents a π-electron donating group) (33). Intermediates of the type HY—B(X)=NR would explain the fast reaction with protic bases of vanishing Brönsted acidity. The results, however, mentioned in Sections V, A, and V, C, favor to some extent the picture of iminoboranes as preferring electrophilic to nucleophilic attack. The high activity of amines can also be rationalized in terms of a concerted process, with a transition state of type **VI**.

VI

C. Boration and Related Reactions

In analogy to the well-known hydroboration, we call addition of $X_2B{-}Cl$, $X_2B{-}N_3$, $X_2B{-}SR$, $X_2B{-}NR_2$, and $X_2B{-}R$ to an unsaturated system a *chloro-*, *azido-*, *thio-*, *amino-*, and *alkyloboration*, respectively.

1. Chloroborations

As a typical dialkyliminoborane, the isopropyl derivative, iPrBNiPr, was chloroborated by three different chloroboranes. In the fast reaction, no alkylo- or aminoboration, respectively, was observed as a potential side reaction competing with the chloroboration [Eq. (30)]. With BCl_3 as chloroborating agent, a vigorous reaction took place which did not yield well-defined products.

$$iPrB{\equiv}NiPr \; + \; {>}B{-}Cl \; \longrightarrow \; \text{(product)} \tag{30}$$

A well-defined product could be isolated from the haloboration of the aminoiminoborane $C_9H_{18}NBNt$Bu with BCl_3 or BBr_3, since the primary product is stabilized by intramolecular BN coordination [Eq. (31a)] (*16*); the corresponding bromoboration yields compound **III**

$$\tag{31}$$

(Section III,B). The same aminoiminoborane undergoes a normal chloroboration with $C_9H_{18}NBCl_2$ [Eq. (31b)] (73).

The aminoiminoborane $Me_3Si—(tBu)N\text{=}B\text{=}NtBu$ can be chloroborated with R_2BCl (R = iBu, sBu) in the expected manner [Eq. (32a)]. The chlorosilane Me_3SiCl is eliminated from the chloroboration product at 140°C [Eq. (32b)], providing a novel synthesis for diazadiboretidines. Addition of chloroboranes R_2BCl with smaller R groups (R = Me, Et, Pr, iPr) proceeds directly to the diazadiboretidine, the primary addition product not being isolable.

$$\text{(32)}$$

Generally, chloroborations seem to be vigorous reactions. The haloboration of 1-alkynes with haloorganoboranes is also known to be rapid under mild conditions (74).

2. Azidoborations

Azidoborations of iminoboranes are smooth and facile reactions [Eq. (33a)] (14, 19). For X = alkyl, the azidoboration products cannot easily be distinguished from hypothetical alkyloboration products

$$\text{(33)}$$

X : Bu iBu $Me_3Si—(tBu)N$

R : tBu iBu tBu

R': Pr   Bu   Bu

[Eq. (33b)], which must be considered, because alkyloboration of iminoboranes with trialkylboranes is a well-known reaction (Section V,C,5). Following addition of [^{10}B]Bu$_2$BN$_3$ to iBuBNiBu, ^{11}B-NMR proved azidoboration to be the actual reaction path. Alkynes cannot be azidoborated.

3. Thioborations

Only one example has been established for thioboration: addition of B(SPr)$_3$ to iPrBNiPr giving 30% yield of PrS—(iPr)B$\stackrel{..}{-}$N(iPr)$\stackrel{..}{-}$B(SPr)$_2$ (19).

4. Aminoborations

The aminoboranes Et$_2$B=NEt$_2$ and Et$_2$B=N(SiMe$_3$)$_2$ do not react with BuB≡NtBu (19). Hydroxylaminoborane derivatives, however, can be brought to reaction with iPrBNiPr [Eq. (34)] (19).

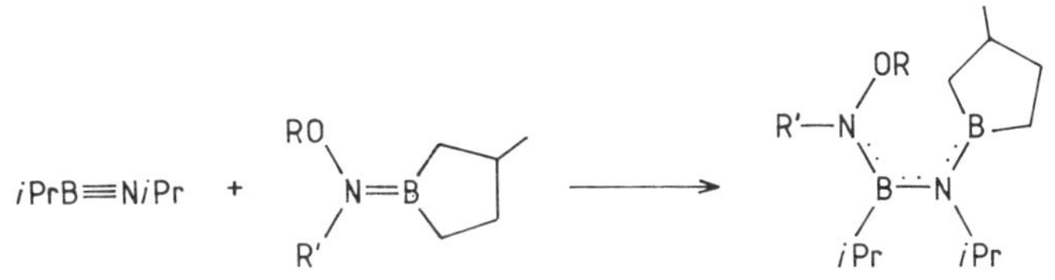

$$i\text{PrB}≡\text{N}i\text{Pr} + \underset{R'}{\overset{RO}{\diagdown}}\text{N=B} \longrightarrow \quad (34)$$

R: Me   Me   SiMe$_3$

R': Me   SiMe$_3$ SiMe$_3$

The same type of reaction was achieved with the hydroxylaminoboranes R$_2$B=N(SiMe$_3$)OSiMe$_3$ and the iminoboranes EtBNEt (R = Pr), iPrBNiPr (R = Bu), BuBNtBu (R = Et) (18, 19). Following the addition of [^{10}B]Bu$_2$B=N(SiMe$_3$)—OSiMe$_3$ to iPrBNiPr, ^{11}B-NMR spectra excluded alkyloboration instead of aminoboration (19).

Hydroxylamino groups differ from normal amino groups by their smaller π-donating power. Hydroxylaminoboranes, therefore, are stronger Lewis acids than aminoboranes, having a vacant boron p-orbital more easily available. The lack of reactivity of aminoboranes indicates that the Lewis acidity of the boranes plays an important role in the boration of iminoboranes. Again, iminoboranes seem to favor electrophilic attack.

5. Organoborations

The ethyloboration of dialkyl iminoboranes (10–13, 17, 18) and aminoiminoboranes (14) with BEt$_3$ is a smooth reaction, that has been applied to nearly all well-characterized iminoboranes [Eq. (35)]. There is not a great difference in going from BEt_3 to BBu_3 (*10*). Phenyloboration with BPh_3 is comparable to alkyloboration (*19*). Looking for trialkylboranes with a larger steric requirement, it turned out that $MeB{\equiv}NMe$ is alkyloborated by $BiPr_3$, but does not react with $BsBu_3$ (*19*).[1]

$$XB{\equiv}NR \ + \ BR'_3 \ \longrightarrow \ \underset{X}{\overset{R'}{}}B{\cdots}N\underset{R}{\overset{BR'_2}{}} \qquad (35)$$

Alkynes $XC{\equiv}CR$ cannot be organoborated, when X represents an alkyl group. In the case of X = H (*74*) or X = Me_3Sn (*75*), a particular type of alkyloboration, coupled to the migration of X, is possible [Eq. (36)]. Such a reaction was not observed with iminoboranes.

$$XC{\equiv}CR \ + \ {>}B{-}R' \ \longrightarrow \ \underset{-B \quad X}{\overset{R' \quad R}{C{=}C}} \qquad (36)$$

6. Allyloboration

The smooth allyloboration of alkynes is known to proceed via an allyl rearrangement, probably including a six-membered cyclic transition state. Thermal treatment of the product initiates a second allyloboration step and a vinyloboration thereafter; the whole procedure [Eq. (37)] opens a synthesis of boraadamantanes by further reaction steps (*76*).

$$XC{\equiv}CH \xrightarrow{+\ B(C_3H_5)_3} \underset{X}{\text{(allyl-B)}} \xrightarrow{40-100^\circ C} \underset{X}{\text{(diene-B)}} {-}C_3H_5 \xrightarrow{130-140^\circ C} X{-}\text{(bicycle)}{-}B{-}C_3H_5 \qquad (37)$$

The allyloboration of the iminoboranes $iPrBNiPr$ and $BuBNtBu$ can readily be accomplished [Eq. (38)]. An intramolecular second allyloboration step requires a temperature of 180°C, demonstrating

---

[1] In this context, *no reaction* means that iminoboranes react faster with themselves than with additional components.

that diallyl(amino)boranes are weaker alloboration reagents than diallyl(vinyl)boranes, which is a consequence of the stronger electron-donating effect of the amino group compared to the vinyl group. The third step in analogy to Eq. (37) would be an intramolecular aminoboration of an olefinic bond. This cannot be achieved, even at 230°C, as was expected, since aminoborations of this type are generally unknown (*19*).

$$R'B\equiv NR \xrightarrow{+B(C_3H_5)_3} \cdots \xrightarrow{180°C} \cdots \xrightarrow{230°C} \cdots \quad (38)$$

7. Addition of Alkylation Reagents

Aminoiminoboranes may be methylated by the esters of strong acids [Eq. (39)]. Benzylation with $PhCH_2Hal$ (Hal = Cl, Br) was not possible (*70*).

$$\cdots \xrightarrow{MeX} \cdots \quad (39)$$

$$X = I, OS(CF_3)O_2$$

8. Addition of Halosilanes and Related Compounds

Iminoboranes R'BNR do not add the silane Me_3SiCl to the $B\equiv N$ triple bond; the synthesis of iminoboranes according to Eq. (1) would then be a reversible and unsuccessful reaction. The aminoiminoborane $C_9H_{18}N\ddot{-}B\dot{=}NtBu$ is reported not to react with Me_3SiCl and Me_3SiBr, but it does react with Me_3SiI and $Me_3Si-OS(CF_3)O_2$ [Eq. (27b)] (Section V,A) (*70*). However, addition of Me_3SiN_3 to the $B\equiv N$ triple bond is a generally applicable reaction, proceeding in analogy to azidoboration [Eq. (33a)]. This "azidosilation" is not a uniform reaction unless the iminoborane is sterically overcrowded [e.g., with $tBuB\equiv NtBu$ (*11*) or $Me_3Si-(tBu)N\ddot{-}B\dot{=}NtBu$ (*14*)]. Usually, a 9:1 mixture of the azidosilation product [Eq. (40a)] and the [3 + 2]-cycloadduct [Eq. (40b)] will be formed (*10–13, 17*).

$$R'B\equiv NR + Me_3SiN_3 \xrightarrow{(a)} \begin{array}{c} N_3 \\ \diagup \\ B=N \\ \diagup \quad \diagdown \\ R' \quad\quad R \end{array} \xrightarrow{SiMe_3}$$

$$\xrightarrow{(b)} \begin{array}{c} Me_3Si \\ \diagdown \\ N \quad N \\ \| \\ N \quad N \\ \diagup \\ B \quad N \\ \diagup \quad \diagdown \\ R' \quad\quad R \end{array}$$

(40)

The reaction of BuB≡N*t*Bu with an excess of Me$_3$SiOEt gives the oxysilation product EtO—(Bu)B—N(*t*Bu)—SiMe$_3$ in good yield. Insofar as cyclic iminoboranes are produced as intermediates, the second of the three products in either of Eqs. (15) or (16) (Section II,B) will apparently have been formed by the addition of Me$_3$Si—OSiMe$_3$ to the reactive intermediate. At first glance, silanes Me$_3$Si—Y will be added to iminoboranes, if the group Y is bonded to silicon via an atom of the second period of the periodic table (e.g., Y = N$_3$, OEt, OSiMe$_3$), whereas addition of Me$_3$SiCl, etc., is not favorable.

D. ADDITION TO BOTH π-BONDS OF IMINOBORANES

In this section, we consider the addition of both A—Y single bonds of AY$_2$ [Eq. (41)], or of an A=Y double bond [Eq. (42)] to the B≡N triple bond of iminoboranes.

$$XB\equiv NR \xrightarrow[+AY_2]{(a)} \begin{array}{c} Y \quad AY \\ \diagdown \diagup \\ B=N \\ \diagup \quad \diagdown \\ X \quad\quad R \end{array} \xrightarrow{(b)} \begin{array}{c} Y \quad A \\ | \quad \| \\ X-B-N \\ | \quad \diagdown \\ Y \quad R \end{array} \xrightarrow{(c)} XBY_2 + A=NR \quad (41)$$

$$XB\equiv NR \xrightarrow[+A=Y]{(a)} \begin{array}{c} Y-A \\ | \quad | \\ B=N \\ \diagup \quad \diagdown \\ X \quad\quad R \end{array} \xrightarrow{(b)} \begin{array}{c} Y \quad A \\ \diagdown \| \\ B-N \\ \diagup \quad \diagdown \\ X \quad\quad R \end{array} \xrightarrow{(c)} \frac{1}{n}(XB=Y)_n + A=NR$$

(42)

There is only one reported example for Eq. (41): addition of WCl_6 to tBuBNtBu gives tBuBCl$_2$ and compound **VII**. Possible intermediates corresponding to hypothetical steps (a) and (b) were not observed (77).

$$\begin{array}{c}
\text{Cl}\text{Cl} \\
t\text{BuN}\!\!=\!\!\overset{|}{\underset{|}{\text{W}}}\!\!-\!\overset{\text{Cl}}{\underset{\text{Cl}}{\diagdown\!\diagup}}\!\!-\!\overset{|}{\underset{|}{\text{W}}}\!\!=\!\!\text{N}t\text{Bu} \\
\text{Cl}\text{Cl} \\
\text{Cl}\text{Cl}
\end{array}$$

VII

Reactions of the type in Eq. (42) have been achieved with aldehydes and ketones [Eq. (43)] (*9, 19*). None of the intermediates according to Eq. (42) was isolated. A [2 + 2]-cycloaddition (step a) probably occurs, since [2 + 2]-cycloadducts can be isolated from the reaction of iminoboranes and oxoalkanes in particular cases (Section VI,A). As far as alkynes are concerned, addition of oxo compounds proceeds in the presence of BF_3 as a catalyst, but without a break of the CC σ-bond [in analogy to step c in Eq. (42)]; rather polar alkynes are needed [Eq. (44)] (*78*).

$$XB\!\equiv\!NR + O\!=\!CR'R'' \longrightarrow \tfrac{1}{3}(XBO)_3 + R'R''C\!=\!NR \quad (43)$$

$$\begin{array}{llllll}
X: & i\text{Pr} & i\text{Pr} & t\text{Bu} & t\text{Bu} & F_5C_6 \\
R: & i\text{Pr} & i\text{Pr} & t\text{Bu} & t\text{Bu} & t\text{Bu} \\
R': & H & H & H & H & Ph \\
R'': & t\text{Bu} & Ph & t\text{Bu} & Ph & Ph
\end{array}$$

$$XC\!\equiv\!CR + O\!=\!C\!\!\diagup \longrightarrow O\!=\!C(X)\!-\!C(R)\!=\!C\!\!\diagup \quad (44)$$
$$X = R'O, R'S, R'_2N, Ph$$

VI. Iminoboranes as Components in Cycloaddition Reactions

A. [2 + 2]-Cycloadditions

Thermal or catalytic cyclodimerization of iminoboranes is obviously a [2 + 2]-cycloaddition (Section IV). A mixture of two different iminoboranes may be stabilized by formation of three different cyclodimers. If the relative stability of the two iminoboranes, however, differs distinctly, the mixed cyclodimer will be formed preferentially by dropping the cooled, less stable component to the relatively warm, more stable one [e.g. Eq. (45)] (*14*).

$$iPr-B\equiv N-iPr \ + \ \underset{tBu}{\overset{Me_3Si}{N}}-B\equiv N-tBu \ \longrightarrow \ \text{(cyclic product)} \quad (45)$$

Reaction of aldehydes and ketones with iminoboranes has been widely investigated. Conditions for the [2 + 2]-cycloaddition between XBNR and R'R"CO are relatively good stability of the iminoborane and lack of enolic protons in the oxo compound [Eq. (46)] (*14, 19*). Relatively less stable iminoboranes, but in some cases the stable ones too, may react with oxo compounds by a total opening of the B≡N triple bond [Eq. (43)], presumably via a [2 + 2]-cycloaddition [Eq. (42)] (Section V,D). A relatively stable iminoborane and a ketone containing enolic protons may yield an open-chain product, probably through a six-membered cyclic transition state [Eq. (46b)] (*19*).

$$XB\equiv NR \ + \ O=CR'R'' \ \xrightarrow{(a)/(b)} \ \text{products} \quad (46)$$

X	:	tBu	Me₃Si(tBu)N	Me₃Si(tBu)N	Me₃Si(tBu)N	tBu	tBu
R	:	tBu	tBu	tBu	tBu	tBu	tBu
R'	:	CF₃	H	H	H	tBu	Ph
R"	:	CF₃	Ph	—CH=CHMe	—CMe=CH₂	Me	Me
Pathway:		a	a	a	a	b	b

Alternative reaction pathways, corresponding to Eq. (46), are also observed in the reaction of iminoboranes with iminoalkanes; the ligand $R^\dagger$, bonded to the iminoalkane nitrogen atom, seems to govern the reaction path [Eq. (47)] (*9, 19*). Offering the two C=N double bonds of

$$XB \equiv NR + R^{\dagger}N=CR'R'' \quad \begin{matrix} (a) \\ \longrightarrow \\ \\ (b) \\ \longrightarrow \end{matrix} \quad \begin{matrix} R^{\dagger}\!\!-\!\!N\!\!-\!\!\overset{R'}{\underset{R''}{C}}\!\!-\!\!R'' \\ \overset{|}{B}\!\!-\!\!\overset{|}{N} \\ X \quad R \\ \\ R'\!\!-\!\!C=CH_2 \\ R^{\dagger}N \\ \underset{X}{B}\!\!-\!\!NHR \end{matrix} \qquad (47)$$

X	: *i*Pr	F₅C₆	*i*Pr	*i*Pr
R	: *i*Pr	*t*Bu	*i*Pr	*i*Pr
R'	: Me	Ph	Me	Ph
R''	: Ph	Ph	Me	Me
R†	: Ph	*t*Bu	*i*Pr	*i*Pr
Pathway:	a	a	b	b

$$iPrB \equiv NiPr + Me_2C=N-N=CMe_2 \longrightarrow \begin{matrix} Me \\ Me_2 \\ C \\ iPrN \quad N-N \\ B \quad B-N \\ iPr \quad iPr \quad iPr \end{matrix} \qquad (48)$$

$Me_2C=N-N=CMe_2$ to the attack of iPrB≡NiPr, both possible paths are realized [Eq. (48)] (*19*).

The aminoiminoborane $C_9H_{18}N-B\equiv NtBu$ gives [2+2]-cycloadducts with the heteroallenes $Y=C=Y'$ [Eq. (49)]. The cycloadducts can be transformed thermally as well as photolytically into the novel four-membered ring systems $(C_9H_{18}NBY)_2$ (Y = O, S, Se), according to Eq. (42b) and (42c) (*79*).

$$C_9H_{18}N-B\equiv NtBu + Y=C=Y' \longrightarrow \qquad (49)$$

Y:	O	S	S	Se
Y':	O	O	S	Se

In the combination of moderately reactive oxoalkanes with highly reactive iminoboranes, cyclodimerization of the iminoboranes is preferred to heterodimerization, but neither the diazadiboretidines nor the corresponding borazines are found as final products. According to Eq. (50a), heterocyclic six-membered rings are the products when no enolic protons are available in the oxo component. Otherwise open-chain products can be isolated, according to Eq. (50b). An exception is acetophenone, which reacts in both ways, despite containing enolic protons. Both reactions [Eqs. (50a) and (50b)], can also be achieved by starting with the same well-defined diazadiboretidines, which are assumed to be intermediates in the reaction of iminoboranes with oxo compounds. Both reaction sequences may go through concerted [4 + 2] steps, indicated by their transition states in Eq. (50), but the alternative two-step process via intermediates cannot be ruled out (9, 19).

$$2 \ XB \equiv NR \longrightarrow \begin{array}{c} X \quad R \\ B-N \\ | \quad | \\ N-B \\ R \quad X \end{array} + O=CR'R'' \begin{array}{c} (a) \\ \\ (b) \end{array} \quad (50)$$

X	:	iPr	iPr	Bu	F$_5$C$_6$	F$_5$C$_6$	F$_5$C$_6$	F$_5$C$_6$
R	:	iPr	iPr	tBu	tBu	tBu	tBu	tBu
R'	:	CF$_3$	Me	CF$_3$	H	H	Me	—(CH$_2$)$_5$—
R''	:	CF$_3$	Ph	CF$_3$	—CH=CH$_2$	—CH=CHMe	Me	
Pathway:		a	a	a	a	a	b	b

Iminoboranes and iminophosphanes may be coupled in a (2 + 2)-cycloaddition [Eq. (51)] (19, 80).

$$XB \equiv NR + R'N=P-NR''_2 \longrightarrow \begin{array}{c} R' \quad NR''_2 \\ N-P \\ | \quad | \\ B-N \\ X \quad R \end{array} \quad (51)$$

X :	Bu	Bu	iBu
R :	tBu	tBu	iBu
R' :	tBu	Me$_3$Si	Me$_3$Si
R'' :	iPr	Me$_3$Si	Me$_3$Si

Turning from iminophosphanes to alkylidenophosphanes (phosphaalkenes), the orientation of the [2 + 2]-cycloaddition is inverted, as far as phosphorus is concerned; only one example has been worked out (product **VIII**) (*19*). The phosphaalkyne tBuC≡P does not react with the iminoborane BuB≡NtBu, which instead trimerizes (*19*). An exotic [2 + 2]-cycloaddition is observed when the very reactive titanaethene

<pre>
  tBu   SiMe₃
    \ |
     P—C—SiMe₃   H₂C—TiCp₂   CF₃
     | | | | |
     B=N   B=N   O—C—CF₃
    /   \   /   \ | |
  iPr   iPr   tBu   tBu   C=CH
                                                 /
         VIII   IX   EtO
                                                      X
</pre>

Cp$_2$Ti=CH$_2$ (Cp = cyclopentadienyl) is liberated from a titanacyclobutane primer by thermal cleavage in the presence of an iminoborane (product **IX**) (*81*). Alkynes may also undergo [2 + 2]-cyclodimerizations with unsaturated polar molecules. Rather polar alkynes seem to be favorable, e.g., ethoxyethyne, which can react with hexafluoroacetone to give the rather unstable product **X** (*78*).

B. [3 + 2]-Cycloadditions

Alkynes have been well explored as dipolarophiles in the [3 + 2]-cycloaddition with almost all possible 1,3-dipoles (*78*), whereas the reaction of iminoboranes as dipolarophiles has focused on covalent azides as 1,3-dipoles. Most well-characterized iminoboranes were reacted with phenyl azide, according to Eq. (52) (*11–14, 17, 20*).

$$XB\equiv NR \ + \ PhN_3 \ \longrightarrow \ \begin{array}{c} Ph \diagdown \ _N\diagup ^{N\diagdown} _N \\ B—N \\ \diagup \diagdown \\ X R \end{array} \quad (52)$$

The same type of product was isolated from the reaction of the iminoborane tBuB≡NtBu with 10 different alkyl azides R'N$_3$ (R' = Me, Et, Pr, Bu, iBu, sBu, n-C$_5$H$_{11}$, cyclo-C$_5$H$_9$, cyclo-C$_6$H$_{11}$, PhCH$_2$) (*19*). The azidosilane Me$_3$SiN$_3$ may also behave as a 1,3-dipole [Eq. (40b)], but addition of the SiN bond to iminoboranes [Eq. (40a)] is usually the preferred reaction (Section V,C,8). This is not so when Me$_3$SiN$_3$ is present during the formation of diaryliminoboranes, ArB≡NAr, as intermediates: Both reaction pathways [Eqs. (40a) and (40b)] are followed to an equal extent for Ar = C_6F_5, o-MeC_6H_4; the [3 + 2]-cycloadduct is the only product in the case of Ar = Ph, Mes (*20*). In contrast to dialkyliminoboranes, the diaryliminoboranes MesB≡NMes and $F_5C_6B{\equiv}NC_6F_5$ as intermediates are not azidoborated (Section V,C,2), but rather undergo [3 + 2]-cycloaddition with excess of the generating reactants Mes_2BN_3 and $(F_5C_6)_2BN_3$, respectively, according to Eq. (11) (*20*).

The nitrone PhCH=N(Me)—O was successfully applied as a 1,3-dipole to MesB≡NMes and to Me_3Si-(tBu)N$\cdots$B$\cdots$NtBu [Eq. (53)] (*14, 20*). The rather reactive iminoborane $F_5C_6B{\equiv}NtBu$, however, cyclodimerizes before being attacked by that nitrone, but the nitrone does attack the initially formed cyclodimer [Eq. (54)] (*9*).

$$XB{\equiv}NR \;+\; O{-}N(Me){=}C(H)(Ph) \longrightarrow \text{[cycloadduct]} \quad (53)$$

$$\text{[dimer]} + O{-}N(Me){=}CHPh \longrightarrow \{\text{intermediates}\} \xrightarrow{-tBuN=CHPh} \text{[product]} \quad (54)$$

C. [4 + 2]-CYCLOADDITIONS

Diels–Alder reactions with alkynes as dienophiles have been known for a long time. Iminoboranes, however, will more readily cyclodimerize than react with dienes, and even the cyclodimers are generally superior to dienes in the competition for excess iminoborane. Among many attempts, therefore, only two reactions with iminoboranes as dienophiles have been successful, and in both of them the diene is cyclopentadiene [Eq. (55)] (*9, 14*).

$$XB{\equiv}NtBu \;+\; \text{cyclopentadiene} \longrightarrow \text{[adduct with B=N]} \quad (55)$$

X = F_5C_6, $Me_3Si(tBu)N$

The formation of Dewar borazines from iminoboranes and their cyclodimers is also a [4 + 2]-cycloaddition, whether or not the Dewar borazines are the final products or are rearranged to normal borazines (Section IV,E).

VII. Iminoboranes in the Coordination Sphere of Transition Metals

Alkynes RC≡CR′ may be η^2-bonded to transition metals M (**XI**). More often alkynes occupy a bridging position between two metal atoms, either perpendicular (**XIIa**) or parallel (**XIIb**) to the M—M bond (*82*).

1:1-Coordination compounds between iminoboranes and transition metals corresponding to **XI** have not yet been detected. Stilbene, $C_{14}H_{12}$, cannot be displaced from $Cp_2Mo(C_{14}H_{12})$, and ethene cannot be displaced from $[(C_2H_4)PtCl_2]_2$, by iminoboranes (*19*), whereas alkynes do replace these alkenes (*83, 84*). Ethene in $(C_2H_4)Pt(PPh_3)_2$ can be substituted by the phosphaalkyne tBuC≡P (*85*), but not by the iminoborane tBuB≡NtBu (*19*). There is a parallel situation with the isoelectronic molecules N_2 and CO. Both are well known to form end-on coordination compounds of the type M—N≡N and M—C≡O, but a sideways coordination, comparable to **XI**, is possible only for N_2, not for CO (*82*). So the structural similarity between the isoelectronic couples XCCR/XBNR and N_2/CO (Section III) finds a counterpart in reactivity, as far as the π-bonds are concerned.

In the infant chemistry of iminoboranes only one example of insertion into a bridge position has been found [Eq. (56)] (*86*). That the structure of the product [Eq. (56)] corresponds to structure **XIIa** has been deduced from the CO absorption bands in the IR spectra, which

tBuB≡NtBu + $Co_2(CO)_8$ $\xrightarrow{-2\ CO}$ (56)

nearly coincide with those of the well-characterized analogous compound $(OC)_6Co_2(tBuCCtBu)$ (87).

η^4-Cyclobutadiene metal compounds may be formed by cyclodimerization of alkynes in the ligand sphere of a metal atom [e.g. $(Ph_4C_4)CoCp$ from $PhC\equiv CPh$ and $CoCp_2$] (88). In contrast to the uncomplexed species, η^4-coordinated cyclobutadienes have a square-planar structure. The compound $BuB\equiv NtBu$ was the first iminoborane that cyclodimerized at a transition metal [Eq. (57a)] (49).

$$\begin{array}{c} 2\ BuB\equiv NtBu \xrightarrow{(a)} \\ \\ (BuBNtBu)_2 \xrightarrow{(b)} \end{array} \xrightarrow[-CO\ -C_4H_8O]{+(C_4H_8O)Cr(CO)_5} \quad [\text{structure}] \tag{57}$$

The same iminoborane is thermally stabilized by cyclotrimerization, but may be cyclodimerized by the catalytic aid of $tBuN\equiv C$ (Section III). The cyclodimer $(BuBNtBu)_2$ produces the same product [Eq. (57b)] as for Eq. (57a). Nine further products of the same type, $M[(R'BNR)_2]$, were prepared either from iminoboranes, from their cyclodimers, or from both; the second starting component was either $M(CO)_5(OC_4H_8)$ (M = Cr, Mo, W), or $Fe(CO)_5$ or $CpCo(C_2H_4)_2$, respectively (Table V).

The structures of $(OC)_4Cr[(BuBNtBu)_2]$ (49), $(OC)_4W[(BuBNtBu)_2]$ (89), and $(OC)_3Fe[(PrBNtBu)_2]$ (89) were determined by X-ray methods. The diazadiboretidine ring skeleton is no longer planar; the MB bonds

TABLE V

η^4-Diazadiboretidine Metal Compounds $M[(R'BNR)_2]$

M	R	R'	Synthesis equation No.	^{11}B-NMR [δ (ppm)]	Reference
$(OC)_4Cr$	tBu	Me	(57a,b)	15.1	13
$(OC)_4Cr$	tBu	Et	(57a,b)	16.4	13
$(OC)_4Cr$	tBu	Pr	(57a,b)	15.7	13
$(OC)_4Cr$	tBu	Bu	(57a,b)	16.7	49
$(OC)_4Cr$	iPr	iPr	(57a,b)	17.6	19
$(OC)_4Mo$	tBu	Et	(57b)	20.6	19
$(OC)_4W$	tBu	Et	(57b)	19.0	19
$(OC)_4W$	tBu	Bu	(57b)	19.5	49
$(OC)_3Fe$	tBu	Pr	(57b)	11.2	19
CpCo	tBu	Pr	(57b)	8.0	19

are longer than the MN bonds with a difference of 15 pm (M = Cr), 16 pm (M = W), and 10 pm (M = Fe) (i.e., nearly the difference between the atomic radii of B and N). The coordination figure around Cr and W is a distorted octahedron of four carbon and two nitrogen atoms. The two CO groups above the CCNN plane, hosting the metal, are bent away from the boron atoms, which cap the two NCN octahedral faces closer to the nitrogen atoms. A simple picture of the bonding situation involves six-coordinated bonds from the carbon and nitrogen atoms along the distorted octahedral axes and two back-donating bonds from Cr or W to the boron atoms, fed by metal d-electrons in orbitals of adequate symmetry.

This picture is supported by ^{11}B-NMR data. In uncomplexed diazadiboretidines of the type $(R'BNR)_2$, ^{11}B-NMR shifts are found in the range 42–45 ppm ($Et_2O \cdot BF_3$ external standard). Provided all BN, BC, and NC bonds in the complexed cyclodimer remained normal σ-bonds, the complexation would make the π-electrons no longer available for boron and the NMR signals of the deshielded ^{11}B-atoms would shift downfield to values beyond 60 ppm, typical for sextet boron atoms (*39*), if there was no back-donation from the metal. In fact, there is a remarkable highfield shift (Table V), demonstrating the feedback of electrons to boron. Measuring the metal-to-boron back-donation by such an ^{11}B-NMR highfield shift, the back-donation is strengthened by going from chromium via iron to cobalt, in parallel to an increasing number of d-electrons. The structural and bonding situation, including ^{11}B-NMR highfield shifts, parallels the situation that is met with the well-known η^6-coordination of borazines to metals of the chromium group [e.g., $(OC)_3Cr[(MeBNMe)_3]$] (*90*).

The picture of the nitrogen atoms in diazadiboretidines acting as Lewis base centers is also supported by the formation of a 1:1 coordination compound with $TiCl_4$ [Eq. (58)] (*91*). The ^{11}B-NMR signal of 22.7 ppm indicates a highfield shift, which cannot be due to d-electrons from tetravalent d^0-titanium. X-Ray structural analysis shows that bridging chlorine atoms provide the observed electronic saturation of the boron atoms.

$$(PrBNtBu)_2 \;+\; TiCl_4 \;\longrightarrow\; \begin{array}{c} \text{complex structure} \end{array} \qquad (58)$$

Acknowledgments

The majority of the synthetic work on iminoboranes reported here is documented in the dissertations of T. Thijssen, S. Würtenberg, W. Pieper, T. von Bennigsen-Mackiewicz, R. Truppat, C. von Plotho, E. Schröder, H. Schwan, K. Delpy, and H.-U. Meier, finished at the Technical University of Aachen in the years from 1975 to 1985. The author is greatly indebted to these scientists for their outstanding work. Dr. H. Maisch's help in preparing the manuscript is gratefully acknowledged.

References

1. Glaser, B., and Nöth, H., *Angew. Chem., Int. Ed. Engl.* **24**, 416 (1985).
2. Lory, E. R., and Porter, R. F., *J. Am. Chem. Soc.* **93**, 6301 (1971).
3. Kirk, R. W., and Timms, P. L., *Chem. Commun.*, p. 18 (1967).
4. Pearson, E. F., and McCormick, R. V., *J. Phys. Chem.* **58**, 1619 (1973).
5. Fehlner, T. P., and Turner, D. W., *J. Am. Chem. Soc.* **95**, 7175 (1973).
6. Kirby, C., Kroto, H. W., and Taylor, M. J., *J. Chem. Soc., Chem. Commun.*, p. 19 (1978).
7. Kirby, C., and Kroto, H. W., *J. Mol. Spectrosc.* **83**, 130 (1980).
8. Lory, E. R., and Porter, R. F., *J. Am. Chem. Soc.* **95**, 1766 (1973).
9. Paetzold, P., Richter, A., Thijssen, T., and Würtenberg, S., *Chem. Ber.* **112**, 3811 (1979).
10. Paetzold, P., and von Plotho, C., *Chem. Ber.* **115**, 2819 (1982).
11. Paetzold, P., von Plotho, C., Schmid, G., Boese, R., Schrader, B., Bougeard, D., Pfeiffer, U., Gleiter, R., and Schäfer, W., *Chem. Ber.* **117**, 1989 (1984).
12. Paetzold, P., von Plotho, C., Schmid, G., and Boese, R., *Z. Naturforsch. B Anorg. Chem. Org. Chem.* **39b**, 1069 (1984).
13. Delpy, K., Meier, H.-U., Paetzold, P., and von Plotho, C., *Z. Naturforsch. B Anorg. Chem. Org. Chem.* **39b**, 1696 (1984).
14. Paetzold, P., Schröder, E., Schmid, G., and Boese, R., *Chem. Ber.* **118**, 3205 (1985).
15. Haase, M., and Klingebiel, U., *Angew. Chem., Int. Ed. Engl.* **24**, 324 (1985).
16. Nöth, H., and Weber, S., *Z. Naturforsch. B Anorg. Chem. Org. Chem.* **38b**, 1460 (1983).
17. Meier, H.-U., Paetzold, P., and Schröder, E., *Chem. Ber.* **117**, 1954 (1984).
18. Paetzold, P., and von Bennigsen-Mackiewicz, T., *Chem. Ber.* **114**, 298 (1981).
19. Paetzold, P., and co-workers, unpublished work.
20. Paetzold, P., and Truppat, R., *Chem. Ber.* **116**, 1531 (1983).
21. Pieper, W., Schmitz, D., and Paetzold, P., *Chem. Ber.* **114**, 1801 (1981).
22. Paetzold, P. I., *Fortschr. Chem. Forsch.* **8**, 437 (1967).
23. Armstrong, D. R., and Clark, D. T., *Theor. Chim. Acta* **24**, 307 (1972).
24. Baird, C., and Datta, R. K., *Inorg. Chem.* **11**, 17 (1972).
25. Viehe, H. G., *Angew. Chem., Int. Ed. Engl.* **4**, 746 (1965).
26. Boese, R., unpublished work, Gesamthochschule-Universität Essen.
27. Simonetta, M., and Gavezzotti, A., *in* "The Chemistry of the Carbon–Carbon Triple Bond" (S. Patai, ed.), Part 1, Chapter 1. Wiley, New York, 1978.
28. Clippard, P. H., Hanson, J. C., and Taylor, R. C., *J. Cryst. Mol. Struct.* **1**, 363 (1971).
29. Hess, H., *Acta Crystallogr. Sect. B* **25**, 2338 (1969).
30. Hanic, F., and Subrtova, V., *Acta Crystallogr. Sect. B* **25**, 405 (1969).
31. Geller, S., and Hoard, J. L., *Acta Crystallogr.* **4**, 399 (1951).
32. Bryan, P. S., and Kuczkowski, R. L., *Inorg. Chem.* **10**, 200 (1971).
33. Nöth, H., Staudigl, R., and Wagner, H.-U., *Inorg. Chem.* **21**, 706 (1982).

34. Bullen, G. J., *J. Chem. Soc., Dalton Trans.*, p. 858 (1973).
35. Robiette, A. G., Sheldrick, G. M., and Sheldrick, W. S., *J. Mol. Struct.* **5,** 423 (1970).
36. Tsai, C., and Streib, W. E., *Acta Crystallogr. Sect. B* **26,** 835 (1970).
37. Zettler, F., and Hess, H., *Chem. Ber.* **108,** 2269 (1975).
38. Schmid, G., Boese, R., and Bläser, D., *Z. Naturforsch. B Anorg. Chem. Org. Chem.* **37,** 1230 (1982).
39. Nöth, H., and Wrackmeyer, B., *in* "NMR Basic Principles and Progress" (P. Diehl, E. Fluck, and R. Kosfeld, eds.), Vol. 14. Springer-Verlag, Berlin and New York, 1978.
40. Klaeboe, P., Bougeard, D., Schrader, B., Paetzold, P., and von Plotho, C., *Spectrochim. Acta Part A* **41,** 53 (1985).
41. Crawford, B. L., and Brinkley, S. R., Jr., *J. Chem. Phys.* **9,** 69 (1941).
42. Sutton, L. E. (ed.), "Tables of Interatomic Distances and Configuration in Molecules and Ions," Spec. Publ. Nr. 11, The Chemical Society, London, 1958.
43. Nakamoto, K., "Infrared and Raman Spectra of Inorganic and Coordination Compounds," Third ed., p. 110. Wiley, New York, 1978.
44. Carlier, P., Dubois, J. E., Masclet, P., and Mouvier, G., *J. Electron Spectrosc. Relat. Phenom.* **7,** 55 (1975).
45. Lisy, J. M., and Klemperer, W., *J. Chem. Phys.* **70,** 228 (1979).
46. "The Nomenclature of Boron Compounds," *Inorg. Chem.* **7** (1968).
47. Berthelot, M., *Ann. Chim. Phys.* **9,** 445 (1866).
48. Keim, W., Behr, A., and Röper, M., *in* "Comprehensive Organometallic Chemistry" (G. Wilkinson, ed.), Vol. 8, Chapter 52. Pergamon, Oxford, 1982.
49. Delpy, K., Schmitz, D., and Paetzold, P., *Chem. Ber.* **116,** 2994 (1983).
50. Bally, T., and Masamune, S., *Tetrahedron* **36,** 343 (1980).
51. Maier, G., Pfriem, S., Schäfer, U., and Matusch, R., *Angew. Chem. Int. Ed. Engl.* **17,** 520 (1978).
52. Irngartinger, H., and Nixdorf, M., *Angew. Chem. Int. Ed. Engl.* **22,** 403 (1983).
53. Gompper, R., and Seybold, G., *Angew. Chem. Int. Ed. Engl.* **7,** 824 (1968).
54. Lindner, H. J., and von Gross, B., *Chem. Ber.* **107,** 598 (1974).
55. Hess, H., *Acta Crystallogr. Sect. B* **25,** 2342 (1969).
56. Baird, N. C., *Inorg. Chem.* **12,** 473 (1973).
57. Heilbronner, E., Jones, T. B., Krebs, A., Maier, G., Malsch, K.-D., Pocklington, J., and Schmelzer, A., *J. Am. Chem. Soc.* **102,** 564 (1980).
58. Meller, A., *in* "Gmelin Handbuch der Anorganischen Chemie," New Supplement Series, Vol. 51. Springer-Verlag, Berlin and New York, 1978.
59. Stock, A., and Pohland, E., *Ber. Dtsch. Chem. Ges.* **59,** 2215 (1926).
60. Wiberg, E., and Bolz, A., *Ber. Dtsch. Chem. Ges.* **73,** 209 (1940).
61. Steuer, H., Meller, A., and Elter, G., *J. Organomet. Chem.*, **295,** 1 (1985).
62. van Tamelen, E. E., and Pappas, S. P., *J. Am. Chem. Soc.* **84,** 3789 (1962).
63. van Tamelen, E. E., and Pappas, S. P., *J. Am. Chem. Soc.* **85,** 3297 (1963).
64. Schäfer, W., and Hellmann, H., *Angew. Chem. Int. Ed. Engl.* **6,** 518 (1967).
65. Maier, G., and Schneider, K.-A., *Angew. Chem. Int. Ed. Engl.* **19,** 1022 (1980).
66. Cardillo, M. J., and Bauer, S. M., *J. Am. Chem. Soc.* **92,** 2399 (1970).
67. Turner, H. S., and Warne, R. J., *Proc. Chem. Soc.*, p. 69 (1962).
68. Clarke, P. T., and Powell, H. M., *J. Chem. Soc. B*, p. 1172 (1966).
69. Nöth, H., and Weber, S., *Chem. Ber.* **118,** 2554 (1985).
70. Nöth, H., and Weber, S., *Chem. Ber.* **118,** 2144 (1985).
71. Paetzold, P., von Plotho, C., Schwan, H., and Meier, H.-U., *Z. Naturforsch. B Anorg. Chem. Org. Chem.* **39,** 610 (1984).
72. Brandl, A., and Nöth, H., *Chem. Ber.* **118,** 3759 (1985).

73. Dirschl, F., Nöth, H., and Wagner, W., *J. Chem. Soc., Chem. Comm.*, p. 1533 (1984).
74. Binnewirtz, R.-J., Klingenberger, H., Welte, R., and Paetzold, P., *Chem. Ber.* **116,** 1271 (1983).
75. Menz, G., and Wrackmeyer, B., *Z. Naturforsch. B Anorg. Chem. Org. Chem.* **32,** 1400 (1977).
76. Mikhailov, B. M., *Usp. Khim.* **45,** 1102 (1976).
77. Stahl, K., Weller, F., Dehnicke, K., and Paetzold, P., *Z. Anorg. Allg. Chem.*, **534,** 93 (1986).
78. Fuks, R., and Viehe, H. G., *in* "Chemistry of Acetylenes" (H. G. Viehe, ed.), Chapter 8. Dekker, New York, 1969.
79. Männig, D., Narula, C. K., Nöth, H., and Wietelmann, U., *Chem. Ber.* **118,** 3748 (1985).
80. Paetzold, P., von Plotho, C., Niecke, E., and Rüger, R., *Chem. Ber.* **116,** 1678 (1983).
81. Paetzold, P., Delpy, K., Hughes, R. P., and Herrmann, W. A., *Chem. Ber.* **118,** 1724 (1985).
82. Cotton, F. A., and Wilkinson, G., "Advanced Inorganic Chemistry," 4th Ed. Wiley, New York, 1980.
83. Herberich, G. E., and Okuda, J., *Chem. Ber.* **117,** 3112 (1984).
84. Chatt, J., Guy, R. G., and Duncanson, L. A., *J. Chem. Soc.*, p. 827 (1961).
85. Nixon, J. F., Burckett-St. Laurent, J. C. T. R., Hitchcock, P. B., and Kroto, H. W., *J. Chem. Soc., Chem. Commun.*, p. 1241 (1981).
86. Paetzold, P., and Delpy, K., *Chem. Ber.* **118,** 2552 (1985).
87. Cotton, F. A., Jamerson, J. D., and Stults, B. R., *J. Am. Chem. Soc.* **98,** 1774 (1976).
88. Kooti, M., and Nixon, J. F., *Inorg. Nucl. Chem. Lett.* **9,** 1031 (1973).
89. Schmid, G., and Boese, R., private communication.
90. Werner, H., Prinz, R., and Deckelmann, E., *Chem. Ber.* **102,** 95 (1969).
91. Schmid, G., Kampmann, D., Meyer, W., Boese, R., Paetzold, P., and Delpy, K., *Chem. Ber.* **118,** 2418 (1985).

SYNTHESIS AND REACTIONS OF PHOSPHORUS-RICH SILYLPHOSPHANES

G. FRITZ

Institut für Anorganische Chemie Der Universität,
7500 Karlsruhe 1, Federal Republic of Germany

I. Introduction

After having prepared H_3Si—PH_2 (*1*) from SiH_4 and PH_3, our interests were extended to the reaction possibilities of the P—H and SiH groups, especially the properties of the Si—P bond and the behavior of the free electron pairs. The Si—P bond appeared to be easily cleaved with HBr or alcohol (*2*), with aluminum halides (*3*), BCl_3 (*4*), and other element halides, as well as with acid chlorides (*5*), or $COCl_2$ (*6*). The easy transferability of the phosphane groups to other elements hence became possible and was used in the preparation of element–phosphorus compounds. In organo-substituted silylphosphanes the Si—P bond is the only reactive bond. The reaction with EtI, and the corresponding one with HI, takes place with the formation of Me_3SiI and $(Et_4P)I$ if an excess of EtI is used. During the reaction at $-78°C$ equimolar amounts of the adducts Me_3Si—$PEt_2 \cdot EtI$, and Me_3Si—$PEt_2 \cdot HI$ are formed. These, when warmed, react by cleavage of the Si—P bond (*7*). Similar adducts are formed with $AlCl_3$ and BCl_3, which, when warmed, also react by cleavage of the Si—P bond and by formation of the element–phosphorus bond (*3, 4*). In silylated transition metal complexes the formation of silylphosphonium compounds like $[Me_3SiPMe_3]^+[Co(CO)_4]^-$ or $[(Me_3Si)_2PMe_2]^+[Co(CO)_4]^-$ from $Mw_3SiCo(CO)_4$ and Me_3P, or Me_3Si—PMe_2, respectively, can also be observed (*8*).

The capacity to form phosphonium salts decreases when negatively charged substituents are introduced at the silicon atom. Thus, with F_3Si—PH_2, only a slight tendency to form adducts and to split the Si—P bond is observed (*9*). Silylphosphanes with a PH_2 group like

Me_3SiPH_2 or H_3Si-PH_2 can be metallated with $LiPEt_2$ without splitting the Si—P bond (10).

$$H_3Si-PH_2 + 2LiPEt_2 \longrightarrow H_3Si-PLi_2 + 2HPEt_2$$

Through reaction with CH_3Cl, $H_3Si-PMe_2$ can then be formed. Compounds of the type $H_{3-x}Me_xSi-PHLi$ can be obtained from the reaction of the PH_2-containing derivative with LiPHMe (10), according to

$$H_3Si-PH_2 + LiPHMe \longrightarrow H_3Si-PHLi + MePH_2$$

These undergo disproportionation at room temperature in ether solution according to

$$2H_3Si-PHLi \longrightarrow (H_3Si)_2PLi + LiPH_2$$

Whereas SiH-containing silylphosphanes such as $H_3Si-PEt_2$ react with $LiPEt_2$ by substituting the SiH group, lithium alkyls on the other hand cleave the Si—P bond (11) as shown in the following case.

$$HSi(PEt_2)_3 + LiMe \longrightarrow MeSiH(PEt_2)_2 + LiPEt_2$$

This finding led to the question, how far in multiply silylated silylphosphanes can Si—P bonds be formed from the cleavage reactions. This was investigated on $P(SiMe_3)_3$ (12). At $-40°C$ in THF the reaction with LiBu proceeds practically completely according to:

$$P(SiMe_3)_3 + LiBu \xrightarrow{THF} LiP(SiMe_3)_2\ 2THF + BuSiMe_3$$

$LiP(SiMe_3)_2 \cdot 2THF$ (white crystals) is outstandingly suitable for the transfer of the $P(SiMe_3)_2$ group, and hence for the preparation of various phosphorus–element compounds (13).

$$LiP(SiMe_3)_2 + CH_2Cl_2 \longrightarrow (Me_3Si)_2P-CH_2-P(SiMe_3)_2$$
$$LiP(SiMe_3)_2 + Me_2SiCl_2 \longrightarrow (Me_3Si)_2P-SiMe_2-P(SiMe_3)_2$$
$$LiP(SiMe_3)_2 + Me_2NBCl_2 \longrightarrow (Me_3Si)_2P-B(NMe_2)-P(SiMe_3)_2$$

Therefore our interest focused on $(Me_3Si)_3P$; the first step was to find a suitable approach to prepare this compound. This turned out to be the reaction of white phosphorus with Na/K alloy and then with Me_3SiCl (14). The formation of $(Me_3Si)_3P$ is based upon the phosphide Na_3P. Accordingly, complete cleavage of the P_4 structure by means of the alkali must have occurred. The next step was to decrease the relative amount of the alkali metal in order to prevent the complete cleavage of the P_4 structure as a prerequisite for the formation of phosphorus-rich phosphides. By reaction of the latter with Me_3SiCl it was believed that phosphorus-rich silylphosphanes should become available. Our further investigations confirmed this hypothesis and resulted in the formation, among other compounds, of $(Me_3Si)_3P_7$ (15) and $(Me_2Si)_3P_4$ (16).

But these are not the only products of the reactions. Other phosphorus-rich compounds, [e.g., $(Me_3Si)_4P_{14}$] were obtained. It appeared reasonable to assume a corresponding phosphide to serve as a basis for the formation of $(Me_3Si)_3P_7$. Subsequently, Li_3P_7 became available as one example of these phosphides through the work of M. Baudler on the reaction of diphosphane with $LiPH_2$ (17). Soon thereafter, both Bauder's group (18) and our own research group (19) obtained this phosphide from white phosphorus.

The investigation of the reaction of the silylated diphosphane $P_2(SiMe_3)_4$ in THF with $LiCMe_3$ showed that Li_3P_7 is also formed among other compounds, through a series of complex reactions from the initially formed $Li(Me_3Si)P-P(SiMe_3)_2$ (20).

Our present knowledge of the chemistry of the phosphorus compounds and in particular of the chemistry of the silylphosphanes is not sufficient for a full understanding and explanation of these complicated reactions in every detail. The investigations reviewed in this article were all undertaken to broaden the basic knowledge in this field with the final goal of permitting a full understanding of these reactions.

II. Formation of $P_7(SiMe_3)_3$

In the reaction of white phosphorus with lithium alkyls, poorly soluble phosphides are first formed, and are subsequently degradated by organometallic compounds. In this degradation, unsubstituted phosphides like Li_3P_7 are formed, as well as partially alkylated phosphides such as $LiP_7(CMe_3)_2$, $Li_2P_7(CMe_3)$, $LiP(CMe_3)_2$, and $LiP_4(CMe_3)_3$ (using $LiCMe_3$) (19). The influence of the concentration ratios on the formation of the compounds which result from the reaction of P_4 with

$LiP(SiMe_3)_2$ (21) can be seen in Scheme 1. Li_3P_7 reacts with $P_7(SiMe_3)_3$ according to Scheme 2.

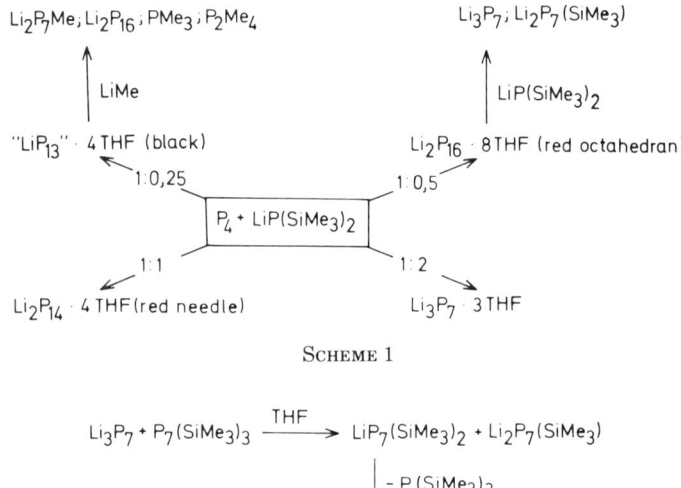

SCHEME 1

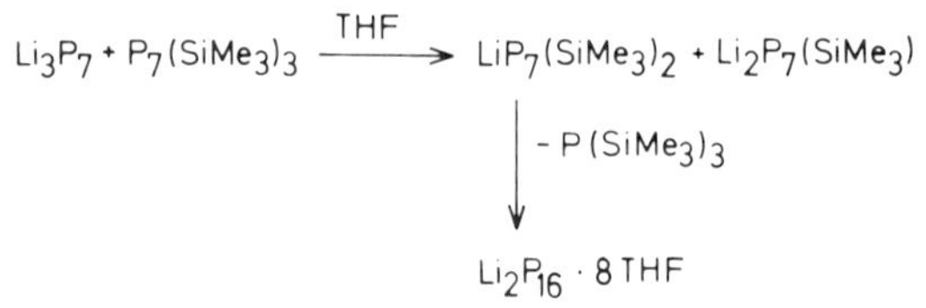

SCHEME 2

$Li_2P_{16} \cdot 8THF$ results from the reaction between P_4 and $LiPH_2 \cdot THF$ as M. Baudler and co-workers (22) were the first to demonstrate. It is also formed in the reaction between P_4 and LiMe or $LiCMe_3$. Whereas $LiP_7(SiMe_3)_2$ yields Li_2P_{16} as $P(SiMe_3)_3$ is split off, $Li_2P_7(SiMe_3)$ reacts to form Li_2P_{14}. The possible pathways for the synthesis of Li_2P_{14} and its reactions are summarized in Scheme 3.

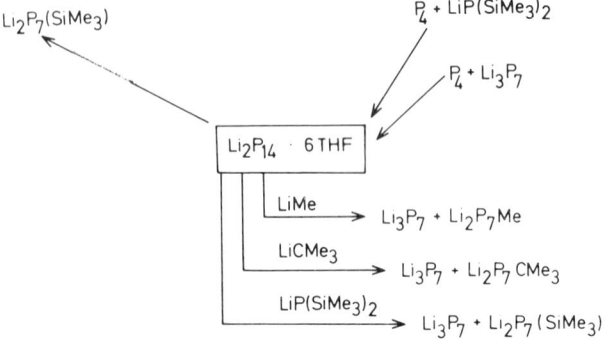

SCHEME 3

Finally, it is possible to obtain phosphorus-rich phosphides by the reaction of P_4 with Li_3P_7:

$$Li_2P_{14}, Li_2P_{16} \xleftarrow{1:1} \boxed{P_4 + Li_3P_7} \xrightarrow{4:1} Li_2P_{16}, \text{"}LiP_{13}\text{"}$$

These equations are deduced from the reactions of every phosphide isolated. The interests in the field of P_7 and related systems of the research group of M. Baudler and those of our own, which are similar but complementary, developed initially from different approaches to the problem. To clarify the structure of the phosphides obtained we tried, though without satisfactory results, to form single crystals suitable for X-ray structural analysis. On the other hand the group of M. Baudler was able to determine the structure of Li_3P_7, Li_2P_{16}, Li_2P_{14} in solution by means of the highly sophisticated ^{31}P-NMR spectroscopy.

From the findings presented concerning the reactions of P_4 it follows that the formation of Li_3P_7 and therefore also of $P_7(SiMe_3)_3$ takes place in several interrelated reaction steps which influence one another, but which cannot yet be detailed. We lack both a detailed picture of the reactive behavior of these compounds and a reliable knowledge of the first steps of the formation reactions. Therefore we sought to understand the problem by investigating simpler systems able to help in its solution. A report of this approach now follows.

III. Formation of Cyclic Silylphosphanes

The reactions of alkali phosphides with R_2SiCl_2 opened the way to the chemistry of the cyclic silylphosphanes. Thus Parshall and Lindsey (23) were the first to report the formation of $Et_2Si(PR)_2SiEt_2$ (R = H, Ph, $SiMe_3$) by reaction of the corresponding lithium phosphide with Et_2SiCl_2, or with the bicyclic compound $P(SiEt_2)_3P$. Schumann and Benda (24) described the compounds $(PhP-SiPh_2)_3$ and $(PhP-SiPh_2)_2$, and West et al. (25) obtained $(PhP-SiMe_2)_2$ and $(PhP-SiMe_2)_3$ by reacting $KHPPh/K_2PPh$ with Ph_2SiCl_2, or Li_2PPh with Me_2SiCl_2, respectively. From the reaction of K_2PPh with $PhSiCl_3$, Schumann and Benda (26) described the formation of $(PPh)_6(SiPh)_4$, which has an adamantane structure. In clear contrast to these seemingly obvious reactions is the formation of cyclic silylphosphanes by rearranging linear silylphosphanes such as $[(Me_3Si)_2P]_2SiMe_2$ to yield the four-membered ring $(Me_3SiP-SiMe_2)_2$ after $(Me_3Si)_3P$ is split off (27), and the preferred formation of $P_4(SiMe_2)_6$ (adamantane structure) as shown in Scheme 4.

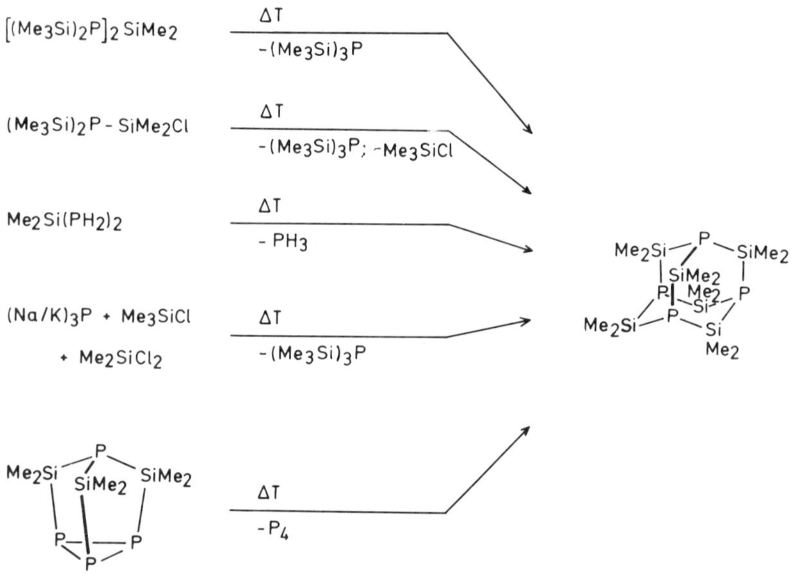

SCHEME 4

Reactions of LiPH(CMe$_3$) with R$_2$SiCl$_2$ (R = Ph, Me, CMe$_3$) lead to (Me$_2$Si—PMe)$_3$, [(Me$_3$C)$_2$Si—PMe]$_2$, Me$_2$Si(P—CMe$_3$)$_2$SiPh$_2$, and the following (28, 29):

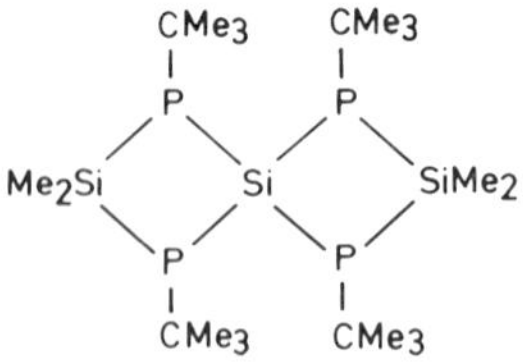

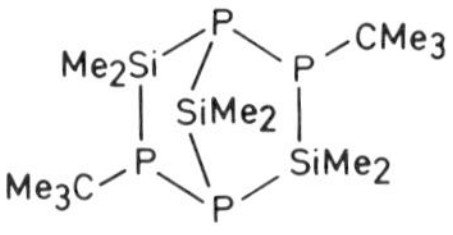

Finally, the reactions of P$_4$ with Na/K alloy and Me$_2$SiCl$_2$ for the preparation of the trisilatetraphosphanortricyclene P$_4$(SiMe$_2$)$_3$, and of (Me$_3$Si)$_3$P$_7$ (15, 16), must be mentioned.

The reactions of lithium phosphides with chlorosilanes, which initially seemed so straightforward, turned out to be strikingly many-sided when PH-containing lithium phosphides are present. The formation of pure LiPH$_2$·DME (DME = 1,2-dimethoxyethane) became feasible by coordination of a high boiling point ether (30). This method was employed by Klingebiel and collaborators (31) for the formation of cyclic phosphanes, according to the following scheme:

$$(t\text{-Bu})_2\text{SiF}_2 \xrightarrow[-\text{LiF}]{+\text{LiPH}_2} (t\text{-Bu})_2\text{Si(F)}-\text{PH}_2 \xrightarrow[-\text{BuH}]{+\text{LiBu}} (t\text{-Bu})_2\text{Si(F)}-\text{PH(Li)}$$

$$\downarrow -\text{LiF}$$

$$\tfrac{1}{2}[\text{HP}-\text{Si}(t\text{-Bu})_2]_2$$

Correspondingly, by reacting R_2SiF_2 with $LiPH(CMe_3)$, these workers produced $[(Me_3C)P-SiR_2]_2$ [R = CMe_3, $-NMe(SiMe_3)$].

A. Reactions of Lithium Phosphides with Me_2SiCl_2

The possible reactions of the lithium phosphides Li_3P, Li_2PH, and $LiPH_2$ with Me_2SiCl_2 are largely determined by the PH groups present. The compounds Li_3P and Na_3P prepared by melting the elements together are well known from the research of Brauer and Zintl (32). Since these substances cannot be used in organometallic reactions because of their low solubility, the metallation of PH_3 with LiBu in ether solution was repeatedly applied to obtain the lithium phosphides. This method was also employed by Parshall and Lindsey to prepare $P(SiEt_2)_3$ (23). But these brownish yellow substances often present several disadvantages: (1) they still contain LiBu, if applied in excess, as well as the Li_3P formed (33); (2) the metallation of PH_3 is not complete; and (3) they consist of a mixture of PH-containing lithium phosphides.

If for the reaction of Me_2SiCl_2 at 0°C a lithium phosphide is used, the latter formed by the introduction of excess PH_3 in $LiBu/hexane/Et_2O$

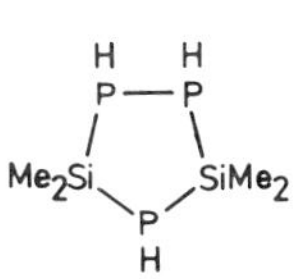

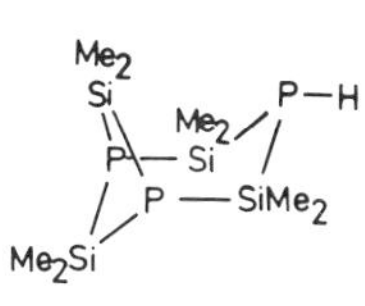

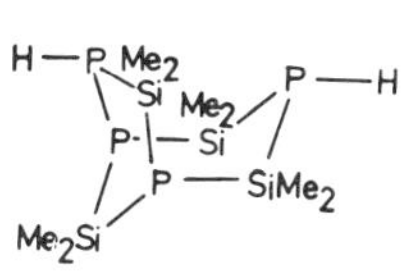

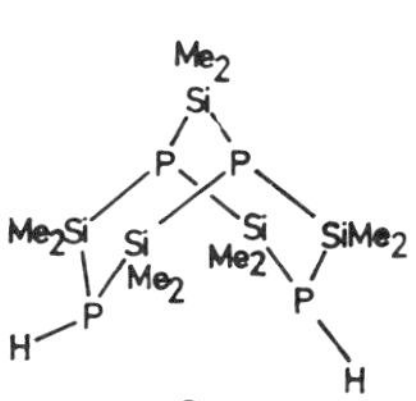

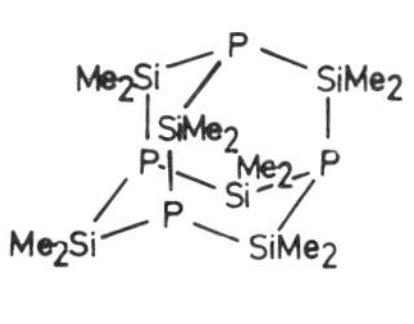

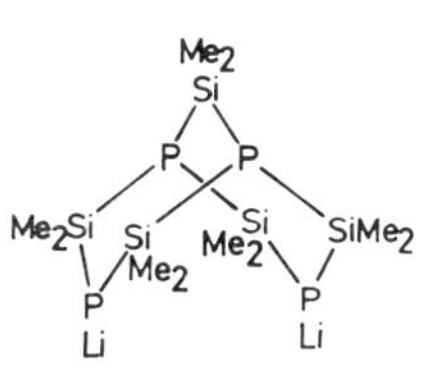

solution (yellow suspension), the reaction products are (34) Me$_2$Si(PH$_2$)Cl (1), Me$_2$Si(PH$_2$)$_2$ (2), H$_2$P—SiMe$_2$—PH—SiMe$_2$Cl (3), (H$_2$P—SiMe$_2$)$_2$PH (4), compound 5, (HP—SiMe$_2$)$_3$ (6), and compounds 7–11; compounds 9 and 10 as well as 6 and 7 are the principal products. The reaction of Me$_2$SiCl$_2$ with excess lithium phosphides produces further metallation of the PH groups and favors formation of polycyclic compounds, as can be perceived by the increase in the concentration of 10 among the reaction products. Besides compound 9 the lithiated derivative 11 is also formed. This can be isolated as an orange-yellow powder and may be transformed by means of Me$_3$SiCl into the silylated compound 12. The bicyclic compound 8 is a by-product of this reaction.

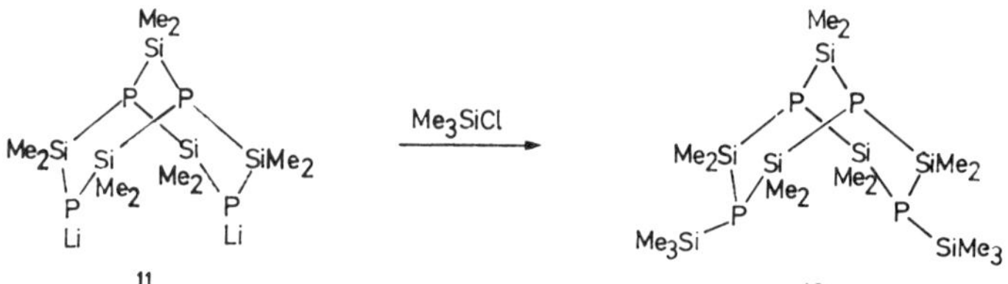

The X-ray structural analysis of compound 9 has been performed (35). Compound 9, a precursor of 10, reacts with (CO)$_4$CrNBD (NBD = 7-nitrobenzo-2-oxa-1,3-diazole) to form the following compound:

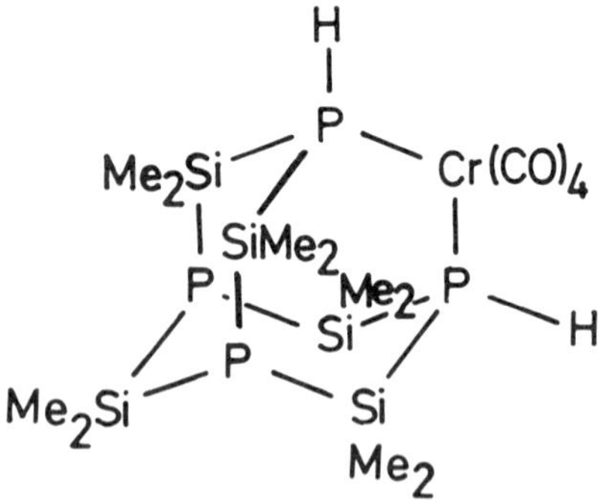

1. Reaction of Li$_2$PH with Me$_2$SiCl$_2$

Nearly pure Li$_2$PH is obtained from the reaction of 1:1 molar proportions of LiPH$_2$·DME and LiBu.

$$\text{LiPH}_2 + \text{LiBu} \longrightarrow \text{Li}_2\text{PH} + \text{BuH}$$

The reaction of Li_2PH in DME leads preferentially to compounds **9** and **10**, which appear as a white powder, from which compound **9** can be obtained as cubic crystals by recrystallization with pentane or toluene. The presence of $Me_2Si(PH_2)Cl$ (**1**), $Me_2Si(PH_2)_2$ (**2**), as well as **3, 4, 5, 6, 7**, and **8** can be demonstrated in the liquid products of the reaction (*34*).

The reaction in pentane with a small admixture of DME progresses considerably slower, but yields, besides compound **10**, principally the same products as obtained in pure DME.

2. Reactions of Li_3P *with* Me_2SiCl_2

It was not possible to obtain Li_3P by metallation of $LiPH_2$ with LiBu due to an ether cleavage reaction. Its formation succeeded through the reaction of PH_3 with an excess of LiBu in a solution of hexane/toluene and through the repeated action of LiBu on the lithium phosphide obtained initially (*34*).

The reaction of this Li_3P with Me_2SiCl_2 in a molar ratio of 1:2 progresses very slowly in DME. After 12 hours at 0°C the suspension of the reaction mixture still has a light brown color. It is only after a 4-hour heating at 84°C that this color slowly clears up. During the reaction, compound **10** (adamantane structure) is formed, as well as small quantities of by-products. There was no indication of the formation of $P(SiMe_2)_3P$.

3. Reaction of $LiPH_2 \cdot DME$ *with* Me_2SiCl_2

In the course of the reaction of $LiPH_2 \cdot DME$ with Me_2SiCl_2 in DME in a molar ratio 2:1 at a temperature of $-40°C$, about 55 mol% of the phosphorus employed is converted to PH_3. The white product that separates contains compound **10** and LiCl. Further concentration of the filtrate causes precipitation of still more compound **10**, accompanied by the evolution of PH_3. The main products of the reaction are PH_3 and compound **10** (*34*).

B. Reaction of Lithium Phosphides with Et_2SiCl_2

The lithium phosphide used (by analogy with Section III,A) reacts with Et_2SiCl_2 in hexane/pentane at 20°C only very slowly. Only after 36 hours does the phosphide suspension lose its color. In addition to LiCl, it was possible to detect compound **13**. On the other hand, formation of compounds **10** and **11** did not take place. By distilling the filtrate the following compounds were separated or at least concentrated in fractions: $Et_2Si(PH_2)Cl$ (**14**) (main product), $Et_2Si(PH_2)_2$ (**15**), $(ClEt_2Si)_2PH$ (**16**), $ClEt_2Si—PH—SiEt_2—PH_2$ (**17**), $(H_2P—SiEt_2)_2PH$ (**18**), $(HP—SiEt_2)_2$ (**19**), and $(HP—SiEt_2)_3$ (**20**). Compounds, **18**, **19**, and **20** could be separated by means of high-pressure liquid chromatography (HPLC) (*34*). Evidence for the presence of $P(SiEt_2)_3P$ was impossible to find. Reaction of the above-mentioned lithium phosphide with Et_2SiCl_2 in Et_2O yielded principally the same compounds, albeit in quantitatively different proportions. In addition, the five-membered ring **21** was observed in small amounts. The compounds **15** (30%) and **18** (10%) are formed as principal products; these were already present in the reaction mixture. Yet neither the four-membered ring **19** nor the six-membered ring **20** is formed. These can first be found in the fractions separated by means of distillation, in which the quantity of **20** then amounts to ~10% of the total silylphosphanes obtained. Derivatives of compounds **10** and **11**, as well as $P(SiEt_2)_3P$, are not formed (*34*).

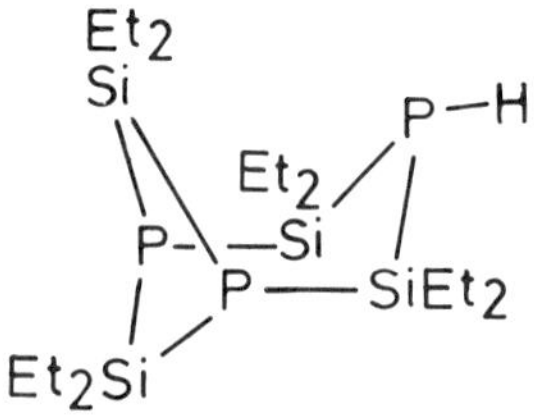

[Structures: compound **13** (left) — bicyclic structure with Et₂Si, Et₂Si, SiEt₂ and P—H groups; compound **21** (right) — five-membered ring HP—PH / Et₂Si, SiEt₂ / P-H]

The reaction of Li_3P (formed from PH_3 and LiBu) with Et_2SiCl_2 in a molar ratio of 2:3 in ether at room temperature progresses only very slowly. After 1 hour compound **13** can be detected among the compounds formed. The reaction in toluene at 20°C shows only little progress after 3 days; after heating at 110°C during 8 hours, the following compounds are formed:

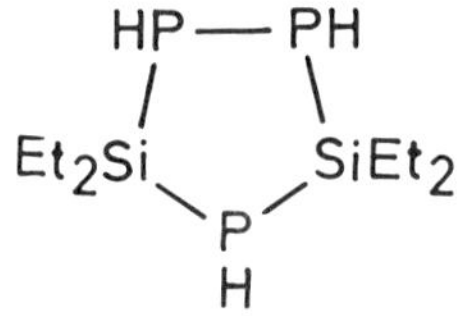

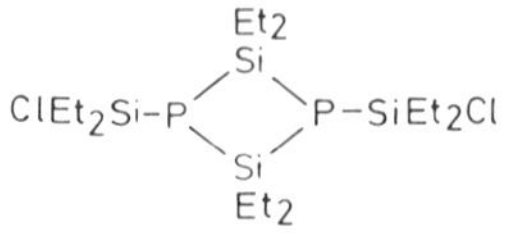

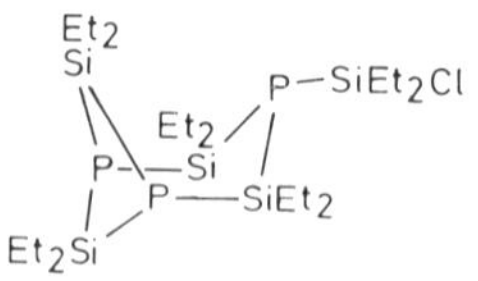

[Structures: left compound with $ClEt_2Si—P$, $P—SiEt_2Cl$, bridged by SiEt₂ groups; compound **22** (right) — bicyclic structure with Et₂Si, Et₂Si, SiEt₂, P—SiEt₂Cl]

In addition, the butylated compounds $ClEt_2SiP(SiEt_2)_2—PSiEt_2Bu$ and $(SiEt_2)_4P_3SiEt_2Bu$ (the butylated derivative of **22**), and trace amounts of **13** are formed. A compound with the adamantane structure corresponding to compound **10** or to its precursor **11** can be excluded with certainty.

C. Reactions of LiPH$_2$·DME and Li$_2$PH with (Me$_3$C)$_2$SiCl$_2$

Introduction of the sterically important Me$_3$C group favors the formation of the four-membered ring. The reaction of (Me$_3$C)$_2$SiCl$_2$ with LiPH$_2$·DME at $-70°$C in a molar ratio of 1.2:2 results in the formation of compound **23**, which was isolated with a yield of 76% in the form of white crystals as shown in Eq. (1). The elimination of PH$_3$

$$4 \text{ LiPH}_2 + 2(\text{Me}_3\text{C})_2\text{SiCl}_2 \longrightarrow \underset{\underset{(\text{CMe}_3)_2}{\text{Si}}}{\overset{\overset{(\text{CMe}_3)_2}{\text{Si}}}{\text{HP}\diamond\text{PH}}} + 2\text{PH}_3 + 4\text{LiCl} \quad (1)$$

$$\underline{23}$$

begins during addition of LiPH$_2$ solution at $-70°$C; 35% of the phosphorus introduced is converted to PH$_3$. This corresponds to the formation of compound **23** with a yield of 70%. At the end there remains a viscous liquid which contains (Me$_3$C)$_2$SiCl$_2$, (Me$_3$C)$_2$Si(PH$_2$)$_2$ (**24**), and (t-Bu)$_2$Si(PH$_2$)Cl (**25**) (*34*).

The reaction of Li$_2$PH with (Me$_3$C)$_2$SiCl$_2$ proceeds very slowly indeed (at 20°C and not before 16 hours have elapsed), to yield compound **23** according to

$$2\text{Li}_2\text{PH} + 2(\text{Me}_3\text{C})_2\text{SiCl}_2 \longrightarrow \text{HP[Si(CMe}_3)_2]_2\text{PH} + 4\text{LiCl}$$

1. Lithiation and Substitution of HP[Si(Me$_3$)$_2$]$_2$PH

The reaction of **23** with LiPH$_2$ in DME yields preferentially compound **26** and PH$_3$. Even with an excess of LiPH$_2$ (molar ratio 1:2) lithiation of the second PH group cannot be achieved. The reaction of HP[Si(CMe$_3$)$_2$]$_2$PH with LiBu in DME leads to the isolation of **26** in the form of colorless crystals (reaction yield 40%), which are found in the reaction mixture together with unreacted **23**; however, it does not entail the formation of LiP[Si(CMe$_3$)$_2$]$_2$PLi (**27**). Increasing the LiBu concentration merely promotes the cleavage of DME. The best approach to the

$$\underset{\underset{(\text{CMe}_3)_2}{\text{Si}}}{\overset{\overset{(\text{CMe}_3)_2}{\text{Si}}}{\text{HP}\diamond\text{PLi}}} \cdot 2 \text{ DME}$$

$$\underline{26}$$

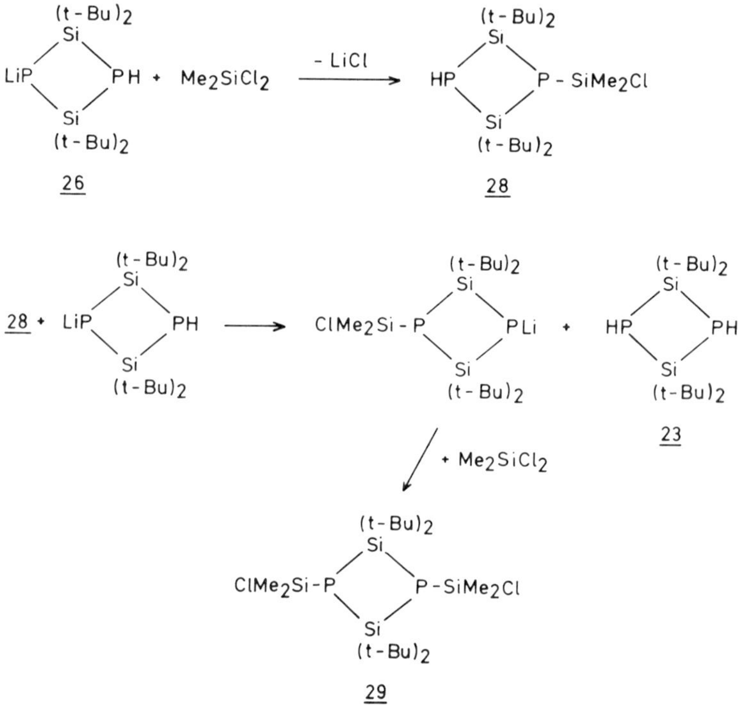

Scheme 5

[Scheme 5 depicting the following reactions:]

$(t-Bu)_2Si$-bridged LiP/PH ring (**26**) + Me_2SiCl_2 $\xrightarrow{-LiCl}$ $(t-Bu)_2Si$-bridged HP/P–SiMe$_2$Cl ring (**28**)

28 + LiP–[$(t-Bu)_2Si$]$_2$–PH $\longrightarrow$ ClMe$_2$Si–P–[$(t-Bu)_2Si$]$_2$–PLi + HP–[$(t-Bu)_2Si$]$_2$–PH (**23**)

$\downarrow$ + Me$_2$SiCl$_2$

ClMe$_2$Si–P–[$(t-Bu)_2Si$]$_2$–P–SiMe$_2$Cl (**29**)

preparation of compound **26** is reaction of **23** with LiCMe$_3$ (molar ratio 1:1 in toluene or pentane/hexane, reaction time 2 days). Since compound **26** is only moderately soluble in these hydrocarbons, it precipitates as a white power (yield 76%). The ether-free compound **26** ignites spontaneously in air and is very soluble in THF. It dissolves in DME if warmed, and it crystallizes out of this solution as LiP[Si(CMe$_3$)$_2$]$_2$PH·2DME. With considerable excess of Me$_2$SiCl$_2$, compound **26** forms, at $-70°$C in DME, a mixture of the following compounds: HP[Si(CMe$_3$)$_2$]$_2$P—SiMe$_2$Cl (**28**) (83%), ClMe$_2$Si—P[Si(CMe$_3$)$_2$]$_2$P—SiMe$_2$Cl (**29**) (10%), and HP[Si(CMe$_3$)$_2$]$_2$PH (**23**) (7%) (percentages are obtained by integrating the ^{31}P-NMR spectra).

Formation of **29** and **23** is due to the transmetallation reaction shown in Scheme 5. By means of fractional crystallization it has been possible to isolate compound **29** in the form of colorless crystals. If the solution obtained by the metallation of **23** with LiBu at $-70°$C is added to Me$_2$SiCl$_2$ at $-50°$C, compound **30** is obtained as shown in Eq. (2).

$$\text{2 LiP} \underset{\underset{(CMe_3)_2}{Si}}{\overset{\overset{(CMe_3)_2}{Si}}{\diamond}} \text{PH} \quad + \text{ Me}_2\text{SiCl}_2 \xrightarrow[-2\text{LiCl}]{-50°\text{C}}$$

$$\text{HP} \underset{\underset{(CMe_3)_2}{Si}}{\overset{\overset{(CMe_3)_2}{Si}}{\diamond}} \text{P - SiMe}_2 \text{- P} \underset{\underset{(CMe_3)_2}{Si}}{\overset{\overset{(CMe_3)_2}{Si}}{\diamond}} \text{PH}$$

30

(2)

Compound **30** separates as a snow white glistening powder. It dissolves well in THF and toluene, poorly in DME and benzene. By metallation of the PH group in compound **28** with LiBu it is not possible to obtain an additional ring closure by elimination of LiCl to yield the corresponding bicyclic molecule. Undetermined molecular associations occur instead.

Reactions of **23** with LiCMe$_3$ in a molar ratio 1:4 in pentane at 20°C (reaction time 24 hours) indicate that metallation of the second PH group in **26** to yield LiP[Si(CMe$_3$)$_2$]$_2$PLi (**27**) is possible. A white solid is formed in this reaction, and the consumption of LiCMe$_3$ corresponds to a double lithiation of compound **23**. There are signs that the dilithiated compound **27** is characterized in the ^{31}P-NMR spectrum by $\delta = -280$ ppm, but it has been impossible to obtain compound **27** in a pure form despite repeated metallation reactions.

The reaction of **23** with LiCMe$_3$ in pentane in a 1:4 molar ratio produces a phosphide, which, besides compound **27**, still contains some **26**. The latter can be further metallated by repeated reactions with LiBu so that an enriched end product is obtained whose content of compound **27** is about 80% (*36*).

The reaction of a lithiated product (free of compound **23**) with Me$_3$SiCl yields compounds **31** (62%), **32** (31%), and **23** (7%). Compound **23** is obtained by transmetallation.

$$\underset{\textbf{31}}{\text{Me}_3\text{Si-P}\underset{\underset{(\text{CMe}_3)_2}{\text{Si}}}{\overset{\overset{(\text{CMe}_3)_2}{\text{Si}}}{\diamond}}\text{P-SiMe}_3} \qquad \underset{\textbf{32}}{\text{HP}\underset{\underset{(\text{CMe}_3)_2}{\text{Si}}}{\overset{\overset{(\text{CMe}_3)_2}{\text{Si}}}{\diamond}}\text{P-SiMe}_3}$$

2. *Reactions of* LiP[Si(CMe$_3$)$_2$]$_2$PH *and*
LiP[Si(CMe$_3$)$_2$]PLi *with* Me$_3$CPCl$_2$

The mixture of compounds **26** and **27** (about 30% **26**) reacts in hexane with t-BuPCl$_2$ and yields as the principal product compound **33** (yellow,

$$\underset{\textbf{33}}{\text{Cl}(\text{CMe}_3)\text{P-P}\underset{\underset{(\text{CMe}_3)_2}{\text{Si}}}{\overset{\overset{(\text{CMe}_3)_2}{\text{Si}}}{\diamond}}\text{P-P}(\text{CMe}_3)\text{Cl}}$$

needle-shaped crystals, easily soluble in pentane and cyclohexane). It also yields compound **23**. Compound **33** is formed in the cis and the trans configuration. The formation of **23** is due to the transmetallation shown in Eq. (3). The analogous reaction in THF yields the compounds

$$\text{LiP}\diamond\text{PH} + \text{HP}\diamond\text{P-P}(\text{CMe}_3)\text{Cl} \longrightarrow \text{HP}\diamond\text{PH} \qquad (3)$$
$$+ \text{LiP}\diamond\text{P-P}(\text{CMe}_3)\text{Cl}$$

HP[Si(CMe$_3$)$_2$]$_2$P—P(CMe$_3$)—P[Si(CMe$_3$)$_2$]$_2$PH (**36**) (main product), HP[Si(CMe$_3$)$_2$]$_2$P—P(CMe$_3$)H (**37**), HP[Si(CMe$_3$)$_2$]$_2$P—P(CMe$_3$)$_2$ (**38**), as well as Cl(CMe$_3$)P—P[Si(CMe$_3$)$_2$]$_2$P—P(CMe$_3$)Cl (**33**), H(CMe$_3$)P—P[Si(CMe$_3$)$_2$]$_2$P—P(CMe$_3$)Cl (**34**), and H(CMe$_3$)P—P[Si(CMe$_3$)$_2$]$_2$P—P(CMe$_3$)H (**35**).

Formation of compounds **34**, **35**, and **37** is explained by Li/Cl exchange, which results from the incomplete metallation of compound **23**, since in the reaction mixture LiCMe$_3$ persists as metallating reagent. This attacks the compounds having a terminal P(CMe$_3$)Cl group by Li/Cl exchange and generates a P(CMe$_3$)Li group and Me$_3$CCl. The latter reacts by forming P(CMe$_3$)H, Me$_2$C=CH$_2$, and LiCl. Compound **38** results from the reaction of HP[Si(CMe$_3$)$_2$]$_2$-P—P(CMe$_3$)Cl (**39**) with excess LiCMe$_3$ (*36*).

3. The Compound P(SiMe$_2$)$_3$P

In 1959 Parshall and Lindsey published their findings concerning the reactions between Li$_3$P and Et$_2$SiCl$_2$, and reported the formation of the bicyclic compound P(SiEt$_2$)$_3$P (*23*).

This compound, which was not completely described at that time, has never been prepared again nor examined by any research group. We were unsuccessful for a long time. But after we had recognized the difficulties which can occur in the reactions of lithium phosphides with R$_2$SiCl$_2$, we turned back to Li$_3$P prepared from elemental lithium and phosphorus. It was suspended in toluene and at 20°C was stirred during 9 or 10 days together with Me$_2$SiCl$_2$. The bicyclic compound P(SiMe$_2$)$_3$P was formed, along with ClMe$_2$Si—P(SiMe$_2$)$_2$P—SiMe$_2$Cl and trace amounts of P$_4$(SiMe$_2$)$_6$ (adamantane) (*37*) (Scheme 6). The bicyclic compound can be distilled off at 110°C/10^{-3} Torr. It is a

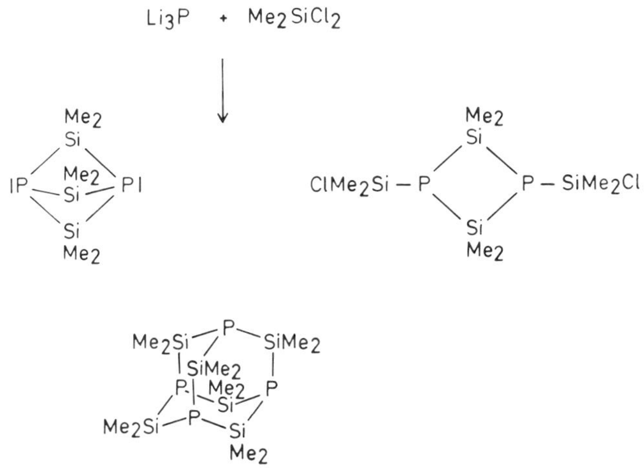

SCHEME 6

liquid, which decomposes to adamantane at a slightly higher temperature [Eq. (4)].

$$2 \begin{array}{c} \text{Me}_2 \\ \text{Si} \\ \diagup \diagdown \\ \text{P} \diagup \text{Me}_2 \diagdown \text{P} \\ \diagdown \text{Si} \diagup \\ \diagdown \diagup \\ \text{Si} \\ \text{Me}_2 \end{array} \longrightarrow \begin{array}{c} \text{Me}_2\text{Si} - \text{P} - \text{SiMe}_2 \\ | \text{SiMe}_2 | \\ \text{P} | \text{Me}_2 \text{P} \\ \diagup - \text{Si} \diagup \\ \text{Me}_2\text{Si} - \text{P} - \text{Si} \\ \text{Me}_2 \end{array} \quad (4)$$

D. Discussion of Results

The investigations performed show that the reactions of the PH-containing phosphides $LiPH_2$ and Li_2PH, or of their mixtures with Li_3P (all prepared from PH_3), with Me_2SiCl_2 lead to compound **10** with an adamantane structure. The intermediate products are a series of cyclic silylphosphanes. Compound **10** is always the main product of the reaction if Me_2SiCl_2 is reacted with an excess of lithium phosphide. For the single reaction steps a transmetallation mechanism can be formulated. This may be recognized both by the evolution of PH_3 during the reactions with $LiPH_2$ and by the isolation of the lithiated compound **11** (formed from compound **9**).

At first the fact that the formation of the bicyclic compound $P(SiMe_2)_3P$ could not be observed seemed incomprehensible. An explanation can be that this compound, under the given reaction conditions, reacts further to form compound **10**. But compound **10** cannot be formed only in this way, as is demonstrated by the isolation of its precursor **9** and of the lithiated compound **11**, which should react with Me_2SiCl_2 to form **10**. Compound **10** is doubtlessly a favored molecule in the series of these cyclic silylphosphanes.

The influence of the substituents on ring size of the reaction products and on the further reactions leading to the formation of polycyclic silylphosphanes is particularly striking. Whereas in the reaction with Me_2SiCl_2 the four-membered ring $(HP-SiMe_2)_2$ was absolutely impossible to obtain and the four-membered ring structure appeared only in the bicyclic compound **7**, in the reactions with Et_2SiCl_2 the four-membered ring $(HP-SiEt_2)_2$ and the six-membered ring $(HP-SiEt_2)_3$ proved to be the favored products. Experimental results indicate that compounds **19** and **20** are formed from the PH-containing linear silylphosphanes only when the reaction products undergo thermal treatment (distillation). In the case of the reation with $(Me_3C)_2SiCl_2$, the main product is $HP[Si(CMe_3)_2]_2PH$ and the six-membered ring no

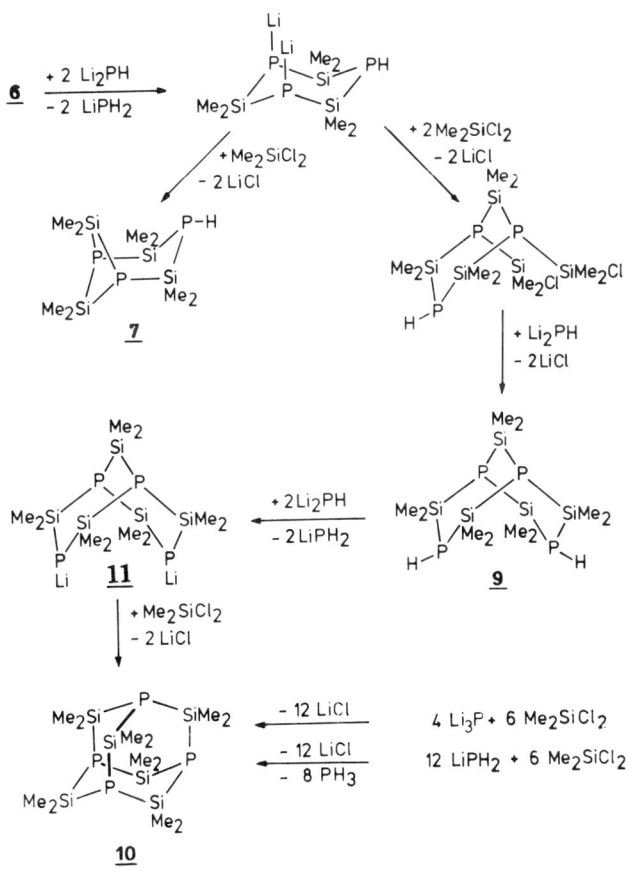

SCHEME 7

longer appears. Compounds **9** and **10**, so strongly favored in the reactions with Me_2SiCl_2, are certainly not formed—under the chosen reaction conditions—in the reactions with Et_2SiCl_2 or $(Me_3C)_2SiCl_2$, respectively. Undoubtedly this depends upon the silicon substituents. Nevertheless, the question remains unanswered whether the failure to format these compounds derives from the deficiency of the corresponding precursor, or whether the formation of an Si—P adamantane with phosphorus atoms in the bridge-head positions and bulky substituents on the Si atoms presents difficulties.

The importance of the intermediate reactions which occur within the overall reaction can be deduced from the multiplicity of the compounds formed during the reaction of $LiPH_2$ or Li_2PH with Me_2SiCl_2. This can, in principle, also be seen in the reaction of $LiPH_2$ with $(Me_3C)_2SiCl_2$ to yield $HP[Si(CMe_3)_2]HP$. The PH_3 obtained can arise by lithiation of a PH-containing intermediate by $LiPH_2$. In this way, $(Me_3C)_2Si(PH_2)_2$, for example, can yield $(Me_3C)_2Si(PHLi)_2$ and PH_3, and react further with $(Me_3C)_2SiCl_2$ to form $HP[Si(CMe_3)_2]_2PH$ and LiCl. The analogous compound $Et_2Si(PH_2)_2$ has also been obtained.

These two reactions, the substitution with elimination of LiCl and the PH lithiation by lithium phosphides, explain the formation of the compounds obtained in the reaction of Me_2SiCl_2 according to the reaction scheme of Scheme 7, in which all identified compounds are given a number.

By metallation of **23** with $LiCMe_3$ the formation of pure $LiP[Si(CMe_3)_2]_2PLi$ (**27**) can never be achieved; instead, a mixture of $HP[Si(CMe_3)_2]_2PLi$ (**26**) and (**27**) (as much as 80%) is obtained. The reactions of these lithiated derivatives of **23** with chloro silanes (Me_2SiCl_2 and Me_3SiCl) or with Me_3CPCl_2 are determined by substitution and transmetallation reactions. These phenomena permit the explanation of the formation of the compounds obtained. Synthesis of a bicyclic compound by further ring closure (e.g., by metallation of **28** and ring closure with elimination of LiCl) is not possible, because the linking to form "chains of rings," as in **30** or **36**, is favored.

IV. Synthesis and Reactions of Silylated Triphosphanes and Triphosphides

$LiP(SiMe_3)_2 \cdot 2THF$ (**12**), which was formed from $(Me_3Si)_3P$ and lithiating agents, proved to be the appropriate reagent for the transfer of the $P(SiMe_3)_2$ group and therefore made possible the preparation of phosphorus functional compounds such as $[(Me_3Si)_2P]_2SiMe_2$. The reaction of $LiP(SiMe_3)_2 \cdot 2THF$ with PCl_3 progresses through

$(Me_3Si)_2P—PCl_2$ to yield $[(Me_3Si)_2P]_2PCl$, which with $LiCMe_3$ forms $[(Me_3Si)_2P]_2PH$; the latter can be lithiated to $[(Me_3Si)_2P]_2PLi$ (38). The chemical behavior of these compounds is evident in nonpolar solvents. In ether solutions the completely silylated phosphanes with two or more phosphorus atoms—as well as their derivatives with PH groups—undergo complicated reactions with lithiating agents. These reactions, after $P(SiMe_3)_3$ and $LiP(SiMe_3)_2$ are eliminated, yield phosphorus-rich phosphides, especially Li_3P_7 (20).

To investigate such reactions further, various substituted, phosphorus-rich silylphosphanes are needed as starting compounds. The following investigations aim at the preparation of phosphorus functional tri- and tetraphosphanes, in which particular phosphorus atoms with reactive substituents (H, Li, halogen, $SiMe_3$) are built in, while other phosphorus atoms with equally defined positions remain blocked by alkyl groups.

A. REACTION OF PCl_3 WITH $(Me_3C)P(SiMe_3)_2$ AND $MeP(SiMe_3)_2$

The reactions of $(Me_3Si)_3P$ with PCl_3 yielding $(Me_3Si)_2P—PCl_2$ indicate, by cleavage of the Si—P bond, with elimination of Me_3SiCl, and by formation of a new P—P bond, a way to construct functional diphosphanes (38). The compounds thus obtained are thermolabile if more Me_3Si-substituted phosphorus atoms are contained in the molecules. The reactions of $(Me_3C)P(SiMe_3)_2$ (41) and $MeP(SiMe_3)_2$ (42) with PCl_3 occur in an analogous manner (20), as shown in Eq. (5) (39).

$$PCl_3 + (Me_3C)P(SiMe_3)_2 \xrightarrow[n\text{-pentane}]{-78°C} (Me_3C)(Me_3Si)P—PCl_2 + Me_3SiCl \quad (5)$$
(40)　　(41)　　　　　　　　　　(43)

Similarly, $MeP(SiMe_3)_2$ yields $Me(Me_3Si)P—PCl_2$ (44). The reaction given in Eq. (5), when set up in 1:1 molar ratio at $-78°C$, produces within 24 hours compound 43 almost quantitatively. The colorless reaction mixture remains apparently unchanged. Compound 43 is unstable in solution. Upon warming at $20°C$ its decomposition occurs with formation of a yellow color.

$MeP(SiMe_3)_2$ (42) reacts with PCl_3 according to Eq. (5) substantially faster than compound 41 does. The reaction in pentane at $-78°C$ is already completed after 15 minutes with the formation of $Me(Me_3Si)$-P—PCl_2 (44). If the concentration of 42 is increased, it can, according to the molar ratio of the reagents, also lead to substitution of the second and third chlorine atoms of PCl_3. Within a few minutes at 20°C **44** decomposes, while the solution takes on an intense yellow color and a yellow, amorphous precipitate is formed.

B. Reaction of $(Me_3C)(Me_3Si)P-PCl_2$ with $LiP(SiMe_3)_2$ and $LiCMe_3$

The reaction of PCl_3 with **41** in both the molar ratios 1:1 and 1:2 produces **43**. No further elimination of Me_3SiCl occurs between **43** and **41** at −78°C in pentane. If, however, **43** is reacted with the phosphides $LiP(SiMe_3)_2$ or $LiP(SiMe_3)CMe_3$, the triphosphanes **46** and **47**, respectively, are formed according to Eq. (6) (*39*).

$$\mathbf{43} + LiP(SiMe_3)R \longrightarrow (Me_3C)(Me_3Si)P-P(Cl)-P(SiMe_3)R + LiCl \qquad (6)$$

$$[R = SiMe_3\ (\mathbf{46}),\ R = CMe_3\ (\mathbf{47})]$$

In order to carry out the reaction, compound **43** must be brought to −78°C and then a solution of $LiP(SiMe_3)_2 \cdot 2THF$ in toluene must be slowly added. Initially the reaction mixture turns yellow and then, because of the separation of LiCl, becomes cloudy. After 2 hours, compound **43** has completely reacted. The main product is $(Me_3C)(Me_3Si)P-P(Cl)-P(SiMe_3)_2$ (**46**) (about 55 mol%), in addition to which about 10% of $[(Me_3Si)_2P]_2PCl$ (**48**) is formed, as well as 20% $(Me_3Si)_3P$, 10% **41**, and 1% $P_2(SiMe_3)_4$ (**49**). Formation of **48** results from compound **46** by cleavage of the P—P bond by $LiP(SiMe_3)_2$ according to Eq. (7).

$$(Me_3C)(Me_3Si)P-P(Cl)-P(SiMe_3)_2 + LiP(SiMe_3)_2 \xrightarrow[n\text{-pentane}]{-78°C}$$
$$\qquad\qquad (\mathbf{46}) \qquad\qquad\qquad\qquad (\mathbf{45})$$

$$\qquad\qquad\qquad (Me_3Si)_2P-P(Cl)-P(SiMe_3)_2 + LiP(SiMe_3)CMe_3 \qquad (7)$$
$$\qquad\qquad\qquad\qquad (\mathbf{48}) \qquad\qquad\qquad\qquad (\mathbf{50})$$

This is even more evident if $LiP(SiMe_3)_2$ is used in excess. Compound **50** and $LiP(SiMe_3)_2$ also, to a small extent, finally react with Me_3SiCl to form $(Me_3Si)_3P$ and **41**. Transmetallation reactions, which by means of an Li/Cl exchange between $LiP(SiMe_3)_2$ and **46** or **43** [intermediate formation of $(Me_3Si)_2PCl$ with $LiP(SiMe_3)_2$ and further reactions] yield the diphosphane **49**, play only a secondary role.

If in order to prepare the triphosphane according to Eq. (6), the analogous bromine compound $(Me_3C)(Me_3Si)P-PBr_2$ is used instead of compound **43**, transmetallation dominates and compound **49** becomes the principal product. The symmetric diphosphane $P_2(SiMe_3)_2(CMe_3)_2$ (**52**) [formed just like **49** from **50** according to Eq. (7)], as well as $P_2(SiMe_3)_3(CMe_3)$ (**51**) (cross product of **50** and **45**), can also be detected, $(Me_3Si)_2P1\text{-}P2(Cl)\text{-}P3(SiMe_3)CMe_3$ contains in P2 and P3 two chiral centers and exists in two diastereomeric configurations which occur in a 6:1 ratio. In solution, **46** is stable only at low temperatures. This can be explained by the presence of both Me_3Si and Cl substituents in the same phosphane molecule. At 20°C it decomposes yielding soluble products which are not easily defined.

C. $(Me_3Si)_2P\text{---}P(H)\text{---}P(SiMe_3)CMe_3$

If the solution of compound **46** obtained according to Eq. (6) is mixed slowly at $-78°C$ with the equivalent amount of $LiCMe_3$ in *n*-pentane, a deepening of the color from yellow to orange is observed. After the reaction mixture is warmed for some hours at 20°C, its color becomes paler. A reaction between $LiCMe_3$ and Me_3SiCl, which according to Eq. (5) is present in the mixture, cannot, under the chosen conditions, be assumed. The principal product of the reaction (with 1:1 molar ratio) is $(Me_3Si)_2P\text{---}P(H)\text{---}P(SiMe_3)CMe_3$ (**53**). It is formed according to Eq. (8) by transmetallation between **46** and $LiCMe_3$.

$$46 + LiCMe_3 \xrightarrow[n\text{-pentane}]{-78°C} (Me_3Si)_2P\text{---}P(Li)\text{---}P(SiMe_3)CMe_3 + Me_3CCl \quad (8)$$
$$(54)$$

Formation of the phosphide **54** can be recognized by the orange color in the reaction solution when $LiCMe_3$ is added. Addition of Me_3CCl causes the subsequent hydrogen substitution for lithium in **54**. As LiCl and isobutene are eliminated, the hydrogenated triphosphane **53** is formed [Eq. (9a)].

$$54 \xrightarrow[-LiCl]{+Me_3CCl} (Me_3Si)_2P\text{---}P(H)\text{---}P(SiMe_3)CMe_3 + H_2C\text{=}CMe_2 \quad (9a)$$
$$(53)$$

$$54 \xrightarrow[-LiCl]{+Me_3SiCl} (Me_3Si)_2P\text{---}P(SiMe_3)\text{---}P(SiMe_3)CMe_3 \quad (9b)$$
$$(55)$$

Simultaneously, **54** forms with Me_3SiCl [from the first reaction stage according to Eq. (5)] the silylated triphosphane **55** [Eq. (9b)]. Substitution by a CMe_3 group, which yields $(Me_3Si)_2P\text{---}P(CMe_3)\text{---}P(SiMe_3)CMe_3$ (**56**), progresses less favorably than silylation of the secondary phosphorus in compounds **54** and **46**, respectively. This is explained by the steric differences between the CMe_3 and the $SiMe_3$ groups.

The triphosphanes $(Me_3Si)_2P-P(H)-P(SiMe_3)_2$ (**57**) and $(Me_3Si)_2P-P(SiMe_3)-P(SiMe_3)_2$ (**58**) are formed from **48** in a manner similar to that in which **53** and **55** are obtained. The main product, **53**, which cannot be separated as a completely pure substance, forms colorless octahedral crystals.

According to the principle stated in Section IV,B,C, by varying the substituents the compounds $[Me_3C)(Me_3Si)P]_2PH$ (**59**), $(Me_3Si)_2$-$P-P(Cl)-P(SiMe_3)Me$ (**60**), $Me(Me_3Si)P-P(Cl)-P(SiMe_3)(CMe_3)$ (**61**), and $[Me(Me_3Si)P]_2PCl$ (**62**) can also be prepared.

D. Preparation and Properties of the Triphosphides $(Me_3Si)_2P-P(Li)-P(SiMe_3)CMe_3$ and $[(Me_3C)(Me_3Si)P]_2PLi$

The phosphides **54** and $[(Me_3C)(Me_3Si)P]_2PLi$ (**63**) can be easily prepared reaction of the derivative hydrogenated on the secondary phosphorus with either n- or t-butyllithium, as Eq. (10) illustrates.

$$R(Me_3Si)P-P(H)-P(SiM3_3)CMe_3 + LiBu \longrightarrow \qquad (10)$$
$$(\textbf{53}, R = SiMe_3; \textbf{59}, R = CMe_3)$$
$$R(Me_3Si)P-P(Li)-P(SiMe_3)CMe_3 + HBu$$
$$(\textbf{54}, R = SiMe_3; \textbf{63}, R = CMe_3)$$

The lithiation shown in Eq. (10) takes places at 20°C in n-pentane. Under these conditions, the reactivity of the lithium alkyl is reduced to such an extent that only the most reactive bond in **53** (and in **59**, respectively) is acted upon and therefore the selective exchange of the phosphorus hydrogen with lithium takes place. The phosphides formed can easily be separated and purified because they dissolve with difficulty in nonpolar solvents. Precipitation of these phosphides is, however, incomplete and can fail to occur if the reaction solutions contain significant concentrations of partially alkylated silylphosphanes such as $(Me_3C)P(SiMe_3)_2$ and $MeP(SiMe_3)_2$.

The ether-free phosphides **54** and $[(Me_3C)(Me_3Si)P]_2PLi$ (**63**) are amorphous pale yellow powders which oxidize spontaneously in air with flames. They can be isolated with a yield of 90–95% if, for their preparation according to Eq. (19), the corresponding hydrogenated triphosphanes **53** and **59**, respectively, are employed in pure form.

If **54** is heated to 133°C, and $[(Me_3C)(Me_3Si)P]_2PLi$ to 194°C, they decompose as indicated by a change in the color of the solution.

E. SUMMARY OF THE RESULTS

The above results show that triphosphanes can be prepared by means of a multiple-stage reaction. They can be obtained experimentally with a one-pot reaction. In the above-quoted examples, the P1 phosphorus atoms are contained in $P(SiMe_3)_2$, $P(SiMe_3)CMe_3$, or $P(SiMe_3)Me$ groups, whereas the P2 phosphorus atoms are substituted with Cl, H, Li, $SiMe_3$, or CMe_3. Formation of the triphosphanes takes place according to the following steps:

1. The first step consists of the reaction of PCl_3 with $P(SiMe_3)_2R$- (R = CMe_3, Me). This reaction at $-78°C$, involving the cleavage of Me_3SiCl, yields the diphosphane $R(Me_3Si)P-PCl_2$ almost quantitatively. In the reactions of PCl_3 with the corresponding phosphides $LiP(SiMe_3)R$ instead of $P(SiMe_3)_2R$, it is not possible to interrupt the reaction at the diphosphane stage. Quite independently from the initial molar ratios, the second chlorine atom in PCl_3 is also substituted, and the triphosphanes $[R(SiMe_3)P]_2PCl$ are formed.

2. The second reaction step requires the use of reactive phosphides:

$$R(Me_3Si)P-PCl_2 + LiP(SiMe_3)R' \longrightarrow R(Me_3Si)P-P(Cl)-P(SiMe_3)R'$$
$$(R = CMe_3, Me; R' = SiMe_3, CMe_3, Me)$$

The main products under all variations of R and R' are the P2-chlorinated triphosphanes. Side reactions such as $SiMe_3/Cl$ exchange, P—P cleavage by $LiP(SiMe_3)R'$, or transmetallations take place only to a small extent. $MeP(SiMe_3)_2$ is an exception inasmuch as it reacts with P-chlorinated phosphanes to yield tri- and even tetraphosphanes.

3. Because of their thermolability (elimination of Me_3SiCl), the P-chlorinated silylphosphanes were reacted with $LiCMe_3$ immediately after their formation at $-78°C$ to yield stable derivatives. In this third reaction step, the following reactions are possible: (A) *tert*-butyl substitution as LiCl is cleaved off; (B) transmetallation yielding phosphides and Me_3CCl: further reactions of the phosphides with Me_3CCl yield P-hydrogenated compounds as isobutene and LiCl are eliminated, and further reactions of the phosphides with Me_3SiCl [from=PCl + $P(SiMe_3)_2R$ as before] lead to the P2-silylated triphosphanes.

Substitution of Cl in $R(Me_3Si)P1-P2(Cl)-P3(SiMe_3)R'$ with the CMe_3 group occurs as the principal reaction only when R = R' = Me. Under all other variations of R and R' the proportion of P2-*t*-butylated compounds remains under 10%. Obviously, for steric reasons, transmetallation between chlorinated triphosphanes and $LiCM_3$ at $-78°C$ is favored (yield 50–60%). The triphosphides formed are nevertheless only intermediate products, which react further with Me_3CCl as well as with Me_3SiCl as the mixture is warmed to room temperature. Two possible complementary parallel reactions occur. With an increasing number of sterically significant substituents R and R' (Me < $SiMe_3$ < CMe_3) in the molecule, H-substitution on P2 (according to Step 3B above) becomes favored relative to silylation as the following comparison shows:

$R(Me_3Si)P1—P2(X)—P3(SiMe_3)R'$		Yield ratio of the triphosphanes with:	
R	R'	X = H	X = $SiMe_3$
$SiMe_3$	CMe_3	3.5	1
Me	CMe_3	1.7	1
Me	$SiMe_3$	0.7	1

In all the reactions investigated, product mixtures are formed which are subjected to a thorough fractionating sublimation. Among the series of compounds $R(Me_3Si)P—P(X)—P(SiMe_3)R'$ formed, the hydrogenated triphosphanes (X = H) can be separated by this means. The products are, if R, R' = $SiMe_3$, **57**; if R = $SiMe_3$, R' = CMe_3, **53**; and if R, R' = CMe_3, **59**. The corresponding silylated derivatives (X = $SiMe_3$) cannot be sublimed; they remain in the residue and undergo thermal decomposition as they reach 60–80°C. With the remaining variations of the substituents R, R' = $SiMe_3$, CMe_3, Me, the corresponding triphosphanes with X = H, $SiMe_3$, CMe_3 can only be enriched in certain sublimed fractions. Lithiation of the hydrogenated triphosphanes **57**, **53**, and **59** is achieved with LiBu almost completely. The resulting phosphides, $[(Me_3Si)_2P]_2PLi$ (**64**), **54**, and **63**, render possible the preparation of further derivatives (X = $SiMe_3$, alkyl, etc.). Furthermore, their reactions in polar ethers are certainly very interesting. interesting.

V. Synthesis of Silylated Tri- and Tetraphosphanes via Lithiated Diphosphanes, and Their Reactions

Synthesis of triphosphanes by Si—P bond cleavage with PCl_3 and further reaction of the triphosphane formed with $LiP(SiMe_3)_2$ is described in Section IV. Starting with the lithium derivative of a suitably substituted diphosphane and linking it to the phosphorus atom of a group with the desired substituents is a method which offered itself for the synthesis of partially silylated and alkylated triphosphanes. A possible route to this synthesis was indicated by the observation that the introduction of one Me$_3$C group in the silylated diphosphane was enough to obtain a lithiated compound stable in ether (40), as shown in the following equation:

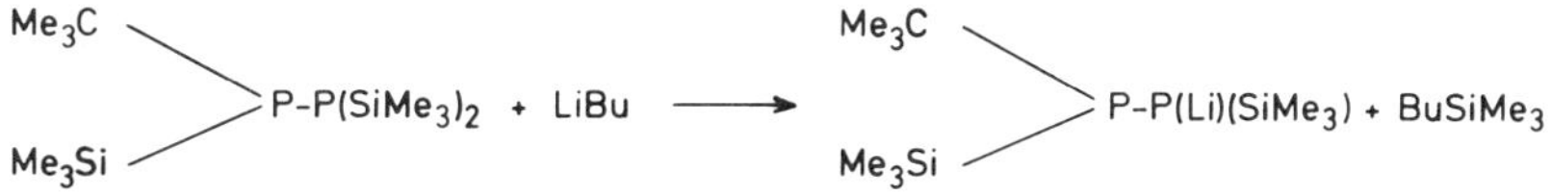

$$\text{Me}_3\text{C}(\text{Me}_3\text{Si})\text{P-P(SiMe}_3)_2 + \text{LiBu} \longrightarrow \text{Me}_3\text{C}(\text{Me}_3\text{Si})\text{P-P(Li)(SiMe}_3) + \text{BuSiMe}_3$$

Scheme 8 gives a further possible variation in the synthesis of functional triphosphanes (40).

Schemes 9 and 10 show the synthesis of triphosphanes with functional substituents in the 1,2-position. Additional reaction proceeds with elimination of Me$_3$SiCl and the formation of the corresponding cyclotetraphosphane in which the P(CMe$_3$)$_2$ groups are adjacent. The reaction progresses with the formation of the P=P double bond, which can be proved by the addition of cyclopentadiene (Scheme 10). The

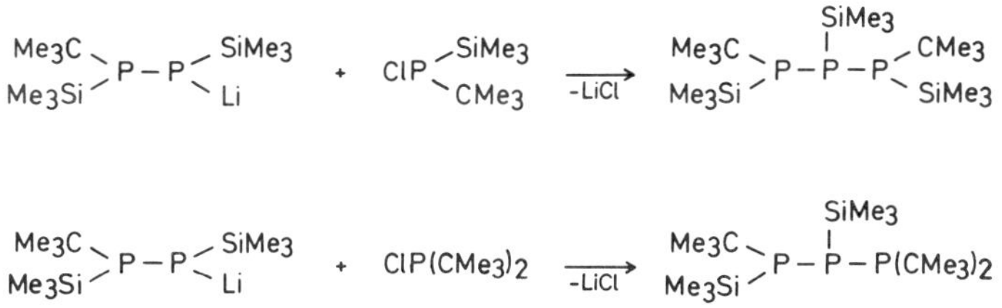

SCHEME 8

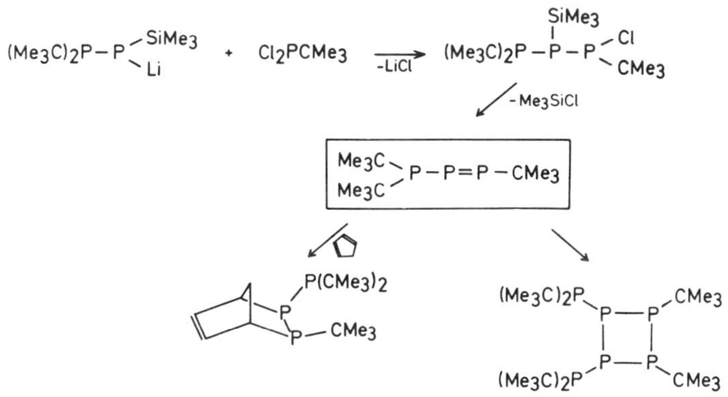

SCHEME 9

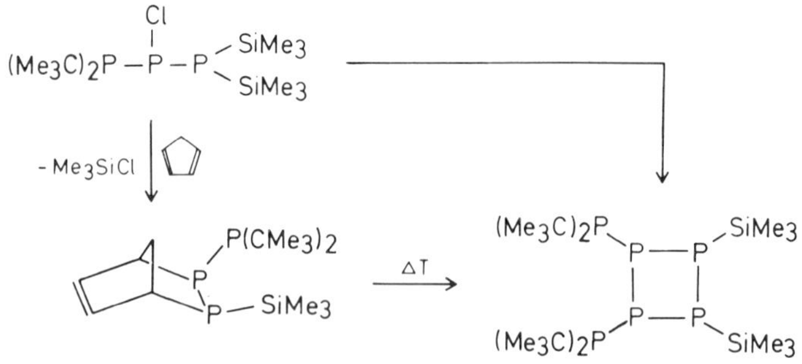

196   G. FRITZ

SCHEME 10

functional groups can also be placed in the 1,3-position of the tetraphosphane, as illustrated by Scheme 11.

The tetraphosphane synthesis developed from the observation by Schumann et al. (41) of the formation of $P_2(SiMe_3)_4$ in the reaction of $LiP(SiMe_3)_2$ with BrH_2C-CH_2Br. The reactions proceed according to Scheme 12.

Scheme 13 illustrates the many reaction possibilities which these compounds exhibit. In Scheme 13a, the reaction of the lithium phosphide with Me_3CPCl_2 results in the formation of the cyclotriphosphane after the elimination of Me_3SiCl from positions 1 and 3. On the other hand, according to Scheme 13b, in the reaction with $(Me_3C)_2PCl$, the lithium phosphide undergoes a lithium exchange reaction. The elimination of Me_3SiCl from the adjacent phosphorus atoms follows; thus the cyclotetraphosphane $P_4[P(CMe_3)_2]_4$ is formed. The resulting Me_3SiCl reacts with $LiP(CMe_3)_2$ (formed by transmetallation) to yield $(Me_3C)_2P-SiMe_3$.

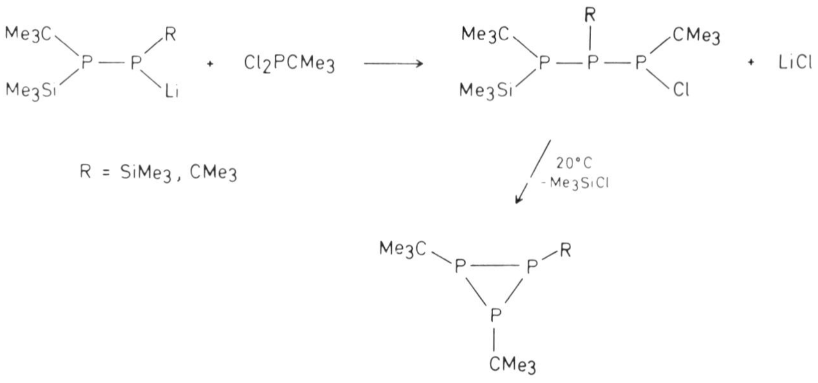

SCHEME 11

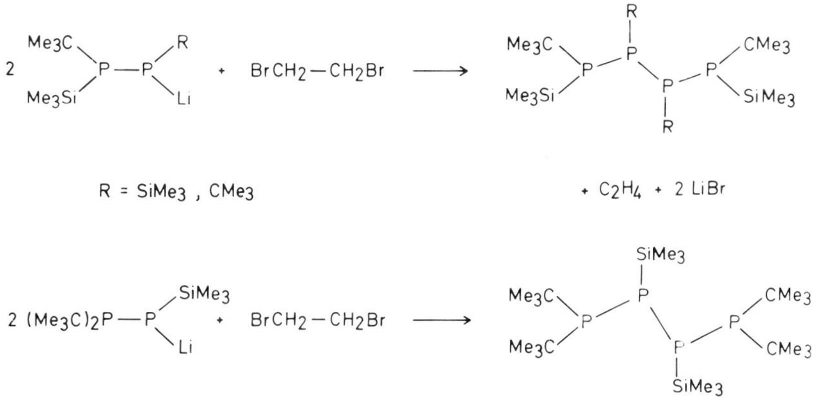

PHOSPHORUS-RICH SILYLPHOSPHANES 197

SCHEME 12

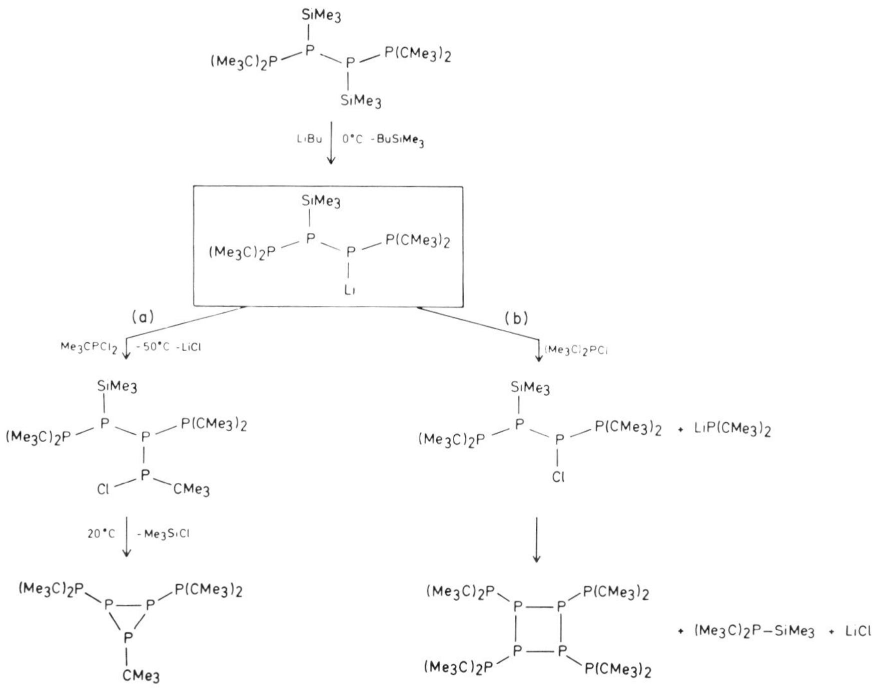

SCHEME 13

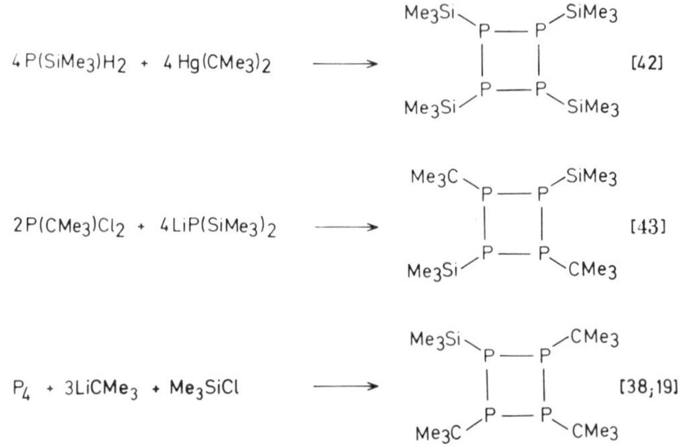

198   G. FRITZ

$$4 P(SiMe_3)H_2 + 4 Hg(CMe_3)_2 \longrightarrow \begin{array}{c} Me_3Si\diagdown_P\diagup P\diagdown^{SiMe_3} \\ | \quad | \\ Me_3Si\diagup^P - P\diagdown_{SiMe_3} \end{array} \quad [42]$$

$$2 P(CMe_3)Cl_2 + 4 LiP(SiMe_3)_2 \longrightarrow \begin{array}{c} Me_3C\diagdown_P\diagup P\diagdown^{SiMe_3} \\ | \quad | \\ Me_3Si\diagup^P - P\diagdown_{CMe_3} \end{array} \quad [43]$$

$$P_4 + 3 LiCMe_3 + Me_3SiCl \longrightarrow \begin{array}{c} Me_3Si\diagdown_P\diagup P\diagdown^{CMe_3} \\ | \quad | \\ Me_3C\diagup^P - P\diagdown_{CMe_3} \end{array} \quad [38;19]$$

SCHEME 14

VI. Silylated Cyclotetraphosphanes

Within the research concerned with the lithiation of silylated, phosphorus-rich phosphanes in ethers, the corresponding cyclotetraphosphanes constitute an interesting group. Those already described in the literature are given in Scheme 14 along with their formation reactions. The missing compounds in this series [i.e., cis-$P_4(CMe_3)_2(SiMe_3)_2$ and $P_4(CMe_3)(SiMe_3)_3$] can be prepared with the aid of the synthesis of the appropriate substituted triphosphanes by eliminating Me_3SiCl, according to Scheme 15. cis-$P_4(CMe_3)_2(SiMe_3)_2$ forms pale yellow crystals of melting point 116°C; $P_4(CMe_3)(SiMe_3)_3$ forms yellow crystals of melting point 143 ± 2°C.

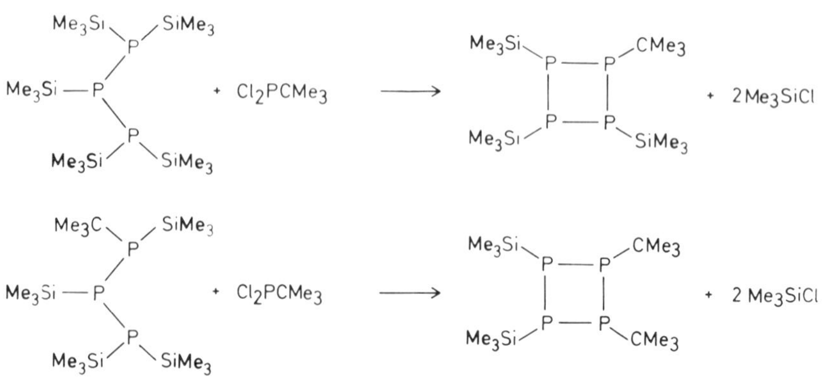

SCHEME 15

VII. Reactions of Silylated Triphosphanes and Cyclotetraphosphanes with Lithium Alkyls

It is known from the preceding research that phosphorus-rich silylphosphanes or their related lithium phosphides undergo, with LiBu in ether, reactions in which a structural transformation occurs, as shown in Scheme 16 (20). The reactions of the partially silylated tri- and cyclotetraphosphanes were explored in order to come closer to understanding the above reactions. It can be taken for granted that the P—C bond is not affected in such reactions.

A. Reactions of Lithium Phosphides $(Me_3Si)_2P$—$P(Li)$—$P(SiMe_3)$ (CMe_3) and $[(Me_3C)(Me_3Si)P]_2PLi$ in Ethers

1. The Reactions of $(Me_3Si)_2P$—$P(Li)$—$P(SiMe_3)(CMe_3)$ (54)

Ether solutions of **54** at $-15°C$ slowly take on an orange color. Over the course of some hours the color becomes deeper, turning wine-red. A day later this intense color becomes paler again and what results is a homogeneous orange-red solution. A steady state is nevertheless reached only after 3–4 days (23°C). These solutions have at this point the same composition as the samples kept for 6 hours at 70°C. The products of the reaction of **54** are the four-membered ring $P_4(CMe_3)_3SiMe_3$(**65**)(38, 39) along with $LiP(SiMe_3)_2$ and $(Me_3Si)_3P$. The compounds Li_3P_7(**66**)(45) and $Li_2P_7CMe_3$(**67**)(19, 46) are formed only in modest amounts, as are the partially alkylated compounds such as $(Me_3C)P(SiMe_3)_2$(**41**), $HP(SiMe_3)CMe_3$(**68**), and $Me_3C(Me_3Si)P$—$P(SiMe_3)Li$(**69**). There exists in other words a parallel between this and the product spectrum of the lithiation of trans-$P_4(CMe_3)_2(SiMe_3)_2$(**71**). Of major (40, 47) significance for the understanding of the development

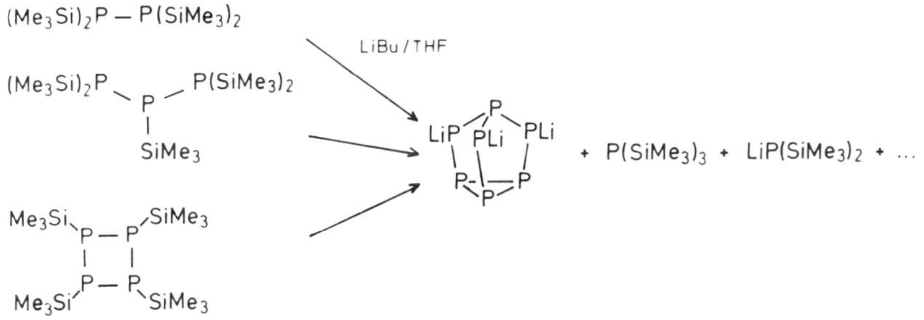

Scheme 16

of the reaction is the formation of the phosphide $(Me_3Si)_2P$—$P(CMe_3)$—$P(SiMe_3)Li$(**72**). This compound, in the early phases of the reaction of $(Me_3Si)_2P$—$P(Li)$—$P(SiMe_3)CMe_3$(**54**), is the first intermediate product which can be detected. This transition from P2 (**54**) to P1 phosphide (**72**) cannot occur by a peripheral exchange of substituents, because under the chosen conditions the P—C bonds are totally inert. Indeed this transition is a sign that the structural rearrangement of the P—P skeleton is already initiated. In the formation of **72** it must be assumed that several intermolecular complex dismutation steps of the P-chain of **54** occur because equimolar amounts of $LiP(SiMe_3)_2$, **68**, and $Li(Me_3Si)P$—$P(SiMe_3)CMe_3$(**73**) are formed, as well as double the amount of $(Me_3Si)_3P$ and traces of $Li(Me_3Si)P$—$P(SiMe_3)_2$(**74**). This implies also the formation of linear-chain tetra- or pentaphosphides. These substances are however unstable under the reaction conditions ($-15°C$, THF) (*44*). They undergo further reactions and thus escape detection by NMR. ^{31}P-NMR examination of the product mixture of the reaction of **54** (after **72** has been formed) yields, at this point in the reaction (homogeneous solution, 4 hours, $-15°C$), the following proportions for the demonstrable atoms and groups: $Li:P:SiMe_3:CMe_3 = 3:9:14:3$. If one compares these values with the triplicated formula of the starting compound **54**, that is $Li_3P_9(SiMe_3)_9(CMe_3)_3$, it is found that the quota of $SiMe_3$ groups in the reaction products is too high. This can be explained by the fact that in the initial homogeneous solution, phosphorus-rich compounds are already present which cannot be detected by ^{31}P-NMR measurements; their signals in the spectrum are masked by the background noise. Their existence can, however, be seen at a more advanced stage of the reaction by a broad unresolved "signal peak," whose chemical shift (80–20 ppm) indicates that unstrained phosphorus five-membered rings probably exist as integral structural elements. Accordingly, in the initial phase of this reaction, through repeated connections of small phosphide units, construction of phosphorus frameworks of higher order is already taking place. At the same time elimination of small molecules and transfer of the silyl substituents to molecules such as $LiP(SiMe_3)_2$ and $(Me_3Si)_3P$ occurs, a fact clearly expressed by the observed concentration ratios of the various groups.

An example of the formation of a ring and of the "desilylation" is given by the formation of *trans*-$P_4(CMe_3)_2(SiMe_3)_2$(**71**) according to Eq. (11), demonstrated, with maximal concentration of compound **72**, after 10 hours at $-15°C$. Closure of the ring to form **71** occurs after $LiP(SiMe_3)_2$ is eliminated, presumably in a two-stage reaction through the *n*-pentaphosphide $Li(Me_3Si)P$—$P(CMe_3)$—$P(SiMe_3)$—$P(CMe_3)$—

P(SiMe$_3$)$_2$, an hypothesized intermediate product which has not yet been detected.

$$2(Me_3Si)_2P - P(CMe_3) - P(SiMe_3)Li \underline{72} \longrightarrow \begin{array}{c} Me_3Si\diagdown P \text{———} P \diagup CMe_3 \\ | \quad\quad\quad | \\ Me_3C \diagup P \text{———} P \diagdown SiMe_3 \\ \underline{71} \end{array} + 2\,LiP(SiMe_3)_2$$

(11)

Formation of **71** is accompanied by metallation by LiP(SiMe$_3$)$_2$ while (Me$_3$Si)$_3$P is eliminated. This means that the further course of the reaction proceeds by the rearrangement of *trans*-LiP$_4$(CMe$_3$)$_2$SiMe$_3$(**70**). The reaction sequence shown raises the P:SiMe$_3$ ratio from the original 1:1 in **72** or **54** to 4:1 in **70**. Due to their property of being excellent leaving groups, the SiMe$_3$ groups are "concentrated" in small molecules such as LiP(SiMe$_3$)$_2$ and (Me$_3$Si)$_3$P, thus causing the formation of new P—P bonds. In the further course of the reaction the P$_4$ ring of **70** reopens and secondary reactions follow, again with the formation of LiP(SiMe$_3$)$_2$ and (Me$_3$Si)$_3$P. The desilylation products are perceived in the ^{31}P-NMR spectrum only by a rather characteristic broad signal peak between 80 and 20 ppm, so that no more information can be obtained about the constitution of the products formed. In this phase of the reaction (3 hours, 23°C) formation of P$_4$(CMe$_3$)$_3$SiMe$_3$ **65** also begins. Once the final stage of the reaction of **54**, after 3 days (23°C), is reached, the proportion of **65** is significantly increased. Coincidently, with a decrease in the intensity of the signal peak the phosphides Li$_3$P$_7$(**66**) and Li$_2$P$_7$CMe$_3$(**67**) can also be detected. The compounds **65**, **66**, and **67** are probably the degradation products of hitherto unknown higher phosphides (*39*).

2. Reactions of [(Me$_3$C)(Me$_3$Si)P]$_2$PLi (63)

Solutions of **63** in THF or DME are stable at 23°C for a few hours. The structure rearranges under these conditions with a half-life of approximately 26 days. For **54**, the half-life at $-15°$C THF is about 6 hours. Without any externally recognizable alteration of the yellow solution of **63**, after a few days, LiP(SiMe$_3$)$_2$ and (Me$_3$C)P(SiMe$_3$)$_2$(**41**) can be observed as initial products in the ^{31}P-NMR spectrum of the solution. These compounds are also formed in the further course of the reaction, always in the same amount. After 60 days (23°C/DME) the solution shows the following composition (values in mol% phosphorus from the integration of the ^{31}P-NMR spectrum): **63**, 14%; LiP(SiMe$_3$)$_2$ (**45**),

26%; **41**, 26%; LiP$_5$(CMe$_3$)$_4$(**76**), 19%; LiP$_3$(CMe$_3$)$_2$(**77**), 5%; LiP$_4$(CMe$_3$)$_3$(**78**), 2%; **69**, 3%; unknown compounds, 5%.

A balance of all the products formed yields the ratio: Li:P:SiMe$_3$:CMe$_3$ = 1:3.2:2:2.2, which, taking into account the accuracy of the measurements is in agreement with the group ratio of the initial compound **69** (39). Accordingly, reaction of **63** yields only "small, defined molecules," whose constitution and relative proportions are easily determined by the ^{31}P-NMR spectra. From the overall equation for the reaction (Scheme 17), the formation of the main products found, LiP(SiMe$_3$)$_2$, (Me$_3$C)P(SiMe$_3$)$_2$(**41**), **76**, **77**, and **78**, can only be qualitatively understood.

As a result, compounds **76** and **77** should be formed in equal proportions. The experimental results differ from this inasmuch as **76** is preferentially formed. This can be explained by the following considerations. It has been shown independently in earlier investigations that the dismutation of **78**, yielding **76** and **77**, is possible in THF. This takes place in the presence of a base [LiBu, LiP(SiMe$_3$)$_2$] even at 0°C. It must be supposed that as precursors of the cyclic phosphides **76**, **77**, and **78** there exist unbranched phosphide chains, whose cyclization with parallel elimination of (Me$_3$C)P(SiMe$_3$)$_2$ yields phosphorus rings of various size according to the length of the chain. The validity of such an assumption is indicated by comparing it to the reaction of cis-P$_4$(CMe$_3$)$_2$(SiMe$_3$)$_2$(**79**) with LiR (R = Me, Bu) in THF (44) (Section VII,B). In that case the first step of the reaction consists exclusively of the opening of the ring of **79**. The resulting n-tetraphosphides Me$_3$C(Me$_3$Si)P—P(CMe$_3$)—P(Li)—P(SiMe$_3$)R and Li(Me$_3$Si)P—P(CMe$_3$)—P(CMe$_3$)—P(SiMe$_3$)R are unstable in THF. Above −30°C

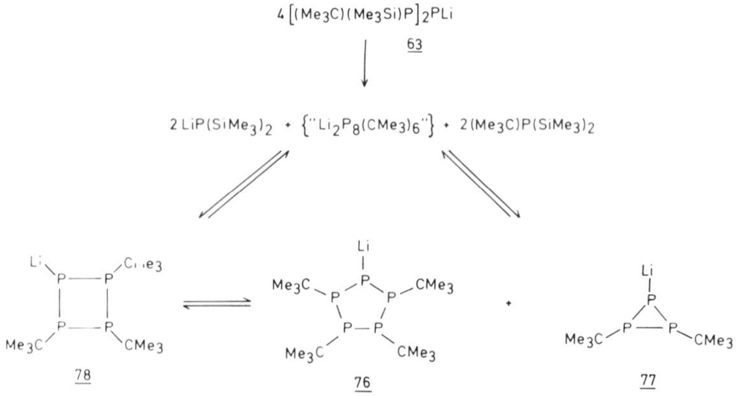

SCHEME 17

they undergo a structural rearrangement which yields among other products **76** and **77**. The analogy with the behavior of **63** becomes evident, if as first reaction step a rearrangement yielding the primary phosphide Li(Me$_3$Si)P—P(CMe$_3$)—P(SiMe$_3$)CMe$_3$(**80**) is postulated, as was observed for the more highly silylated homologues **54** and **64**. An NMR spectroscopic proof of formation of **80** and of further intermediates is not possible at the temperature required for the reaction to take place (23°C), because the secondary reactions proceed faster than the primary step. By elimination of LiP(SiMe$_3$)$_2$, linear-chain higher phosphides can be constructed from **80**. These can nevertheless not be lengthened to any desired extent. Their decreasing stability as their size increases is manifested by dismutations; in these the chain lengths are altered by breaking and reforming P—P bonds in an intermolecular reaction (48–54). The only possibility to stabilize the unbranched phosphides remains ring closure as smaller molecules like **41** or LiP(SiMe$_3$)$_2$ are split off. Ramifications, which can likewise stabilize larger phosphorus structures, can develop only if secondary phosphorus atoms are also substituted in a chain by means of functional groups (rearrangement of **74**, **64**, or **54**).

These considerations permit the compilation of Scheme 18, which summarizes the rearrangement of **63**.

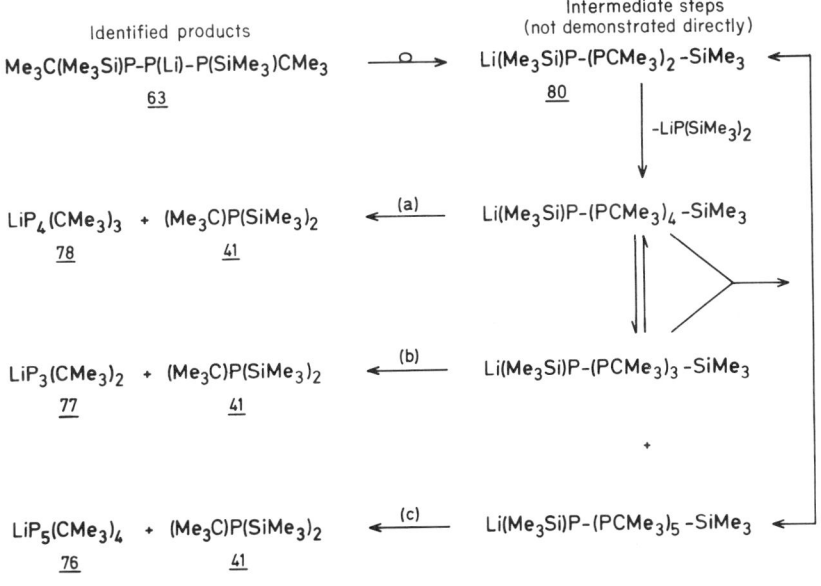

SCHEME 18

The reactions given in Scheme 18 provide a satisfying explanation for the formation of all the observed products and their relative proportions. The observed preferred tendency toward formation of **76** can also be explained. The stability of linear-chain phosphanes is significantly increased by substitution with bulky *tert*-butyl groups. Thus $(PCMe_3)_4(SiMe_3)_2$ (*50*) or $(PCMe_3)_4H_2$ (*50*) can be heated above 60°C without decomposing, while the phenyl compounds $(PC_6H_5)_4(SiMe_3)_2$ and $(PC_6H_5)_4H_2$ (*49*) decompose rapidly even at room temperature. It is therefore plausible that the *tert*-butyl substitution of adjacent phosphorus atoms stabilizes the dismutation of **80** as far as the *n*-hexaphosphide $LiP_6(CMe_3)_5(SiMe_3)_2$, before cyclization to the pentaphosphide **76** occurs (reaction c, Scheme 18). Formation of **77** according to step (b) and that of **78** according to step (a) are consequently only collateral branches of the main reaction. Comparison of the reactions of the compounds $(Me_3Si)_2P$—$P(Li)$—$P(SiMe_3)_2$, $(Me_3Si)_2P$—$P(Li)$—$P(CMe_3)(SiMe_3)$, and $[(Me_3C)(Me_3Si)P]_2PLi$ clearly shows the influence of the Me_3C groups upon the course of the reactions.

B. Reactions of Silylated Cyclotetraphosphanes with Lithium Alkyls

Phosphorus–silicon bonds in trimethylsilyl phosphanes can be cleaved by lithium alkyls (*11*). Such reactions occur in most cases even below 20°C in the presence of a solvating ether like THF or DME.

$$\equiv\!P\!-\!SiMe_3 + LiR \longrightarrow LiP\!\!\equiv + RSiMe_3 \qquad (12)$$

In the series of silylated cyclotetraphosphanes $P_4(CMe_3)_n(SiMe_3)_{4-n}$, ($n = 0\text{–}3$), the behavior of $P_4(CMe_3)_3SiMe_3$ (**65**) and *trans*-$P_4(CMe_3)_2(SiMe_3)_2$ (**71**) toward lithium alkyls [LiMe, Li(n − Bu)] was investigated earlier (*47*). Both compounds react according to Eq. (12). Whereas the phosphide $LiP_4(CMe_3)_3$ (**81**) formed from **65** can be obtained as a crystalline THF adduct, *trans*-$LiP_4(CMe_3)_2SiMe_3$ (**70**) could not be isolated in the past (*40*), because secondary reactions which alter the phosphorus structure of **70** take place during the course of its formation from **71**. Compound **70** yields the following products: the cyclotetraphosphane **65** and $R'P(SiMe_3)_2$ ($R' = Li$, $SiMe_3$, CMe_3, Me, or *n*-Bu), as well as phosphorus-rich unidentified compounds. The formation of the P_7 framework is only observed in the form of $Li_2P_7CMe_3$ and $LiP_7(CMe_3)_2$, and then only to a small extent (<5%).

The reactive behavior of $P_4(SiMe_3)_4$ (**85**) with respect to lithium alkyls was also the object of former investigations (*47*). It is these that firmly established the formation of Li_3P_7, $P(SiMe_3)_3$, and $LiP(SiMe_3)_2$.

1. Reactions of $P_4(SiMe_3)_4$ with Lithium Alkyls

The reactions of **85** with LiR (R = Me, n-Bu) in stoichiometric quantities are completed within a few minutes in THF at a temperature between -60 and $-50°C$. These reactions do not yield the initially expected $LiP_4(SiMe_3)_3$; instead, the secondary phosphide $(Me_3Si)_2P$—$P(SiMe_3)$—$P(Li)$—$P(SiMe_3)R$ is formed almost exclusively. The latter is not the primary product of ring opening. By altering the solvent polarity, the first-formed product can be obtained. If the reaction of **85** with LiR is performed in a solvent mixture in which $Et_2O:THF = 14:1$, at a temperature of between -50 and $-40°C$, after 30 minutes reaction time the formation of the expected primary n-tetraphosphide can be demonstrated (44) [Eq. (13)].

$$P_4(SiMe_3)_4 + LiR \longrightarrow Li(PSiMe_3)_4R \quad (13)$$

(85)   (R = Me, **86**; R = n-Bu, **87**)

In pure Et_2O (without added THF) the reaction times are considerably lengthened. Accordingly, in the first reaction step, a P—P bond of **85** is cleaved by the nucleophilic attacking lithium alkyl. Because of the symmetrical structure of **85** only one primary product can be yielded by the opening of its four-membered ring. When cooled THF is added to this solution of compound **86**, or of **87**, rapid isomerization to the P2 n-tetraphosphide follows, as illustrated by Eq. (14).

$$Li(PSiMe_3)_4R \xrightarrow{THF} (Me_3Si)_2P—P(SiMe_3)—P(Li)—P(SiMe_3)R \quad (14)$$

(R = Me, **86**; R = n-Bu, **87**)   (R = Me, **88**; R = n-Bu, **89**)

The isomerization according to Eq. (14) is in contrast with the behavior of the secondary triphosphides $(Me_3Si)_2P$—$P(Li)$—$P(SiMe_3)R$ (R = $SiMe_3$, CMe_3). These compounds under similar conditions yield the primary phosphides $(Me_3Si)_2P$—$P(R)$—$P(SiMe_3)Li$, whose formation is nevertheless linked with further reactions which change the P framework (20). On the other hand, the 1,3-displacement shown in Eq. (14) between lithium and a functional $SiMe_3$ group takes place quantitatively without the formation of further products. The orange-yellow solutions of the silylated n-tetraphosphides in THF are stable for several hours at $-78°C$. They decompose between -30 and $-20°C$ within a few minutes and at room temperature within seconds with the appearance of a deep red color.

The ^{31}P-NMR spectra of the homogeneous solutions in this phase of the reaction show only the formation of $(Me_3Si)_3P$ and $LiP(SiMe_3)_2$, as well as small quantities of RP(SiMe$_3$)$_2$ (R = Me, n-Bu); information about the fate of the remaining phosphorus is lost in the background noise. Further changes in the composition of the solutions take place only slowly. In accordance with previous investigations (47), at 20°C the formation of Li$_3$P$_7$ can be demonstrated only after several days. Numerous resonance signals of weaker intensity indicate that in the final stage of the reaction of silylated n-tetraphosphides, along with the products mentioned, other phosphorus-rich compounds are also formed, whose identification has not yet been possible (44).

a. *Preparation and Properties of the n-Tetraphosphanes Formed from* P$_4$(SiMe$_3$)$_4$. The thermolabile n-tetraphosphides obtained from **85** and LiMe or Li(n-Bu) have been detected only through their ^{31}P-NMR spectra. These compounds yield with Me$_3$SiCl or MeCl the stable n-tetraphosphanes whose isolation is possible. Thus the results given in Section VII,B,1 are further confirmed. The changes shown in Eqs. (15)–(17) proceed without side reactions at low temperatures with sufficient speed so that, in comparison to them, the structural modifications of the n-tetraphosphide are negligible (44).

For its transformation into the corresponding silyl derivative, the orange-yellow solution of the n-tetraphosphide **88**, immediately after its formation from **85** (with LiMe in THF, −50°C), is reacted with Me$_3$SiCl. The reaction's progress, described in Eq. (15), can be recognized by a change in color to pale yellow.

$$(Me_3Si)_2P-P(SiMe_3)-P(Li)-P(SiMe_3)Me \xrightarrow[-LiCl]{+Me_3SiCl} Me_3Si(PSiMe_3)_4Me \quad (15)$$
$$(88) \qquad\qquad\qquad\qquad\qquad\qquad\qquad (90)$$

After slow warming to 20°C, a clear colorless solution finally results, whose ^{31}P-NMR spectrum indicates nearly complete formation of **90**. In the following work-up, compound **90** can be isolated as colorless needle-shaped crystals. As its melting point is reached (about 49°C), it degrades slowly, manifested by a turning of the color to yellow. At this point, formation of MeP(SiMe$_3$)$_2$ can be demonstrated. Formation of **90** according to Eq. (15) indirectly proves that the reaction of the cyclotetraphosphane **85** with LiMe yields an open-chain n-tetraphosphide; the position occupied by lithium is nevertheless not yet determined. In order to reach a definite conclusion about this, the phosphide **89**, obtained in a further reaction of **85** with Li(n-Bu) (THF/ −50°C), was transformed into the methyl derivative **91** by MeCl, as indicated in Eq. (16).

$(Me_3Si)_2P\text{---}P(SiMe_3)\text{---}P(Li)\text{---}P(SiMe_3)Bu + MeCl \longrightarrow$
(89)

$\qquad\qquad (Me_3Si)_2P\text{---}P(SiMe_3)\text{---}P(Me)\text{---}P(SiMe_3)Bu + LiCl \quad (16)$
$\qquad\qquad\qquad\qquad\qquad (91)$

The formation of **91** shown in Eq. (16) is also manifested by the gradually paler color of the orange-yellow solution of **89**.

As further proof of the course of the reaction between **85** and lithium alkyls, the primary product **86**, resulting from the opening of the ring in compound **85** according to Eq. (13), was reacted anew with MeCl (44). At $-40°C$, formation of the symmetrical n-tetraphosphane **92** takes place, as shown in Eq. (17). This is an equally stable derivative, whose $^{31}P\{^1H\}$-NMR spectrum shows the characteristic resonances of an AA'XX' spin system.

$\qquad\qquad Li(PSiMe_3)_4Me + MeCl \longrightarrow Me(PSiMe_3)_4Me + LiCl \qquad (17)$
$\qquad\qquad\quad (86) \qquad\qquad\qquad\qquad\qquad (92)$

2. Reactions of $P_4(SiMe_3)_3CMe_3$ with Lithium Alkyls

Reactions of $P_4(SiMe_3)_3CMe_3$ with LiR (R = Me, n-Bu) have been conducted under the same conditions as were the reactions of $P_4(SiMe_3)_4$ given in Section VII,B,1. The cyclophosphane reacts completely with LiR at $-45°C$ in THF or DME within a few minutes, whereas in nonpolar hydrocarbons no reaction is observed after weeks even at $20°C$. The reaction consists of the opening of the ring in $P_4(SiMe_3)_3CMe_3$ by P—P bond cleavage, as shown in Scheme 19; linear-chain n-tetraphosphides are formed (44).

The lithium alkyl attacks as a nucleophile the P4 atom of $P_4(CMe_3)(SiMe_3)_3$, so that its bond with P1 or P3 is cleaved. In both cases, the same compounds are formed. The reactions of $P_4(CMe_3)(SiMe_3)_3$ with LiR yield in THF nearly quantitatively the secondary n-tetraphosphides **95** and **96**. Moreover, the primary n-tetraphosphides **93** and **94** occur only as intermediate products and they cannot be detected by ^{31}P-NMR spectroscopy. Their identification can nevertheless be achieved if instead of pure THF a 9:1 Et_2O:THF solvent mixture is employed. As more THF is added at -50 to $-40°C$ to a solution of **93** or **94**, isomerization into the secondary n-tetraphosphides **95** or **96**, respectively, is accomplished by means of a 1,3-displacement between lithium and an $SiMe_3$ group. The reaction pathway given is thus confirmed.

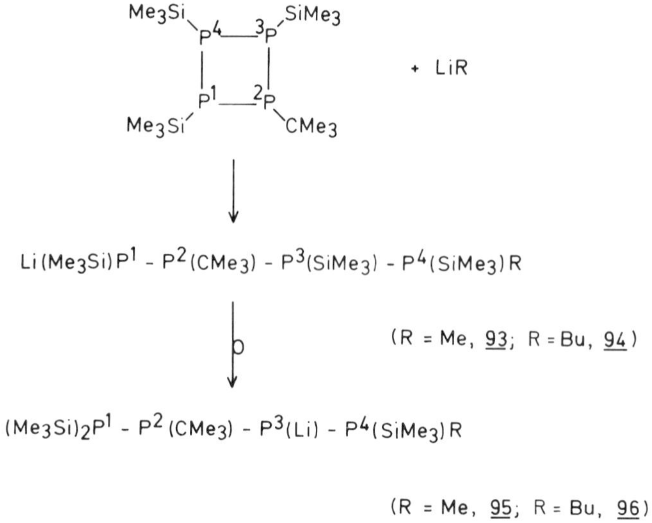

$$\begin{array}{c}\text{Me}_3\text{Si}\diagdown\overset{4}{\text{P}}\text{---}\overset{3}{\text{P}}\diagup\text{SiMe}_3\\|\quad\quad|\\\overset{1}{\text{P}}\text{---}\overset{2}{\text{P}}\\\text{Me}_3\text{Si}\diagup\quad\diagdown\text{CMe}_3\end{array}\quad+\text{LiR}$$

$$\downarrow$$

Li(Me$_3$Si)P^1 - P^2(CMe$_3$) - P^3(SiMe$_3$) - P^4(SiMe$_3$)R

$$\downarrow$$ (R = Me, <u>93</u>; R = Bu, <u>94</u>)

(Me$_3$Si)$_2$P^1 - P^2(CMe$_3$) - P^3(Li) - P^4(SiMe$_3$)R

(R = Me, <u>95</u>; R = Bu, <u>96</u>)

SCHEME 19

The n-tetraphosphides formed according to Scheme 19 are stable for some hours at $-40°C$ in the presence of THF. As these orange-yellow solutions are brought to room temperature a quick decomposition takes place; thereafter the formation of LiP(SiMe$_3$)$_2$, (Me$_3$Si)$_3$P, and minor quantities of RP(SiMe$_3$)$_2$ (R = Me, n-Bu) can be demonstrated. Further compounds cannot be observed even after the reaction solution is allowed to stand several days. After 10–20 hours heating of the NMR samples at 70°C, traces of Li$_3$P$_7$ and of Li$_2$P$_7$CMe$_3$ can be detected. The ^{31}P-NMR spectra measurements of relatively concentrated solutions indicate, with the appropriate amplification, that the bulk of the effective products of the decomposition is concealed in a noncharacteristic broad "signal peak" between +100 and 140 ppm. Further identification of these products by this method is not feasible.

a. Preparation and Properties of the n-Tetraphosphanes from P$_4$(SiMe$_3$)$_3$CMe$_3$. The thermolabile n-tetraphosphides obtained from the reactions of P$_4$(SiMe$_3$)$_3$(Me$_3$) with lithium alkyls have been transformed with Me$_3$SiCl or MeCl into more stable n-tetraphosphanes (*44*). The reactions illustrated in Eq. (18) proceed quickly and without side reactions at about $-40°C$. They generate with high yields the corresponding silyl or methyl derivatives of the phosphides.

It was furthermore possible to prove that, as expected, during the silylation of primary and of secondary n-tetraphosphides, as Eqs. (18b) and (18c) indicate, the same products are formed. The extraordinarily

$(Me_3Si)_2P-P(CMe_3)-P(Me)-P(SiMe_3)R$ $R = Me, (n\text{-}Bu)$ (18a)
   97   (98) $\uparrow$ +MeCl
                      −LiCl

$(Me_3Si)_2P-P(CMe_3)-P(Li)-P(SiMe_3)R$ (18b)
   95   (96) $\downarrow$ +Me$_3$SiCl
                      −LiCl

$(Me_3Si)_2P-P(CMe_3)-P(SiMe_3)-P(SiMe_3)R$ (18c)
   99   (100) $\uparrow$ +Me$_3$SiCl
                       −LiCl

$Li(Me_3Si)P-P(CMe_3)-P(SiMe_3)-P(SiMe_3)R$ (18d)
   93   (94) $\downarrow$ +MeCl
                      −LiCl

$Me(Me_3Si)P-P(CMe_3)-P(SiMe_3)-P(SiMe_3)R$
   101   (102)

high solubility of the *n*-tetraphosphanes and the extreme tendency of their solutions to become supersaturated render their isolation as solid crystals quite difficult. Attempts to sublime such substances led to their thermal decomposition. Therefore, for their characterization by ^{31}P- and ^{1}H-NMR as well as for mass spectrometric investigations, the product mixtures—with the exception of compound **100**—as shown in Eq. (18) were used. In these mixtures the respective *n*-tetraphosphanes were present in proportions of 75–95%.

The multiple-step reaction of $P_4(SiMe_3)_3CMe_3$ with Li(*n*-Bu) and Me$_3$SiCl shown in Scheme 19 and in Eq. (18b) generate the *n*-tetraphosphane **100** with a 95% yield. During the final work-up, **100** is obtained initially as a colorless, highly viscous oil, which can occasionally become light yellow because of slight oxidation. By crystallization in a little *n*-pentane, colorless, rod-shaped crystals are obtained, which melt at 83 ± 3°C and gradually decompose. Compared with the cyclic tetraphosphane $P_4(SiMe_3)_3CMe_3$, the P—Si and P—P bonds in the *n*-tetraphosphane **100** differ distinctly in their behavior with Li(*n*-Bu). Whereas in $P_4(SiMe_3)_3CMe_3$ Li(*n*-Bu) (THF, −45°C) cleaves only a P—P bond and thus alters the phosphorus molecular skeleton, under comparable reaction conditions Li(*n*-Bu) cleaves an Si—P bond in compound **100**. As a result, *n*-BuSiMe$_3$ is eliminated and **96** is formed. This formation would seem, for steric and statistical reasons, unfavorable, but in terms of the fast isomerization of the primary *n*-tetraphosphide **94** into the secondary product **96** [Scheme 19, step (b)], this can be completely understood.

C. Reactions of cis-$P_4(SiMe_3)_2(CMe_3)_2$ (79) with n-Butyllithium

The cyclotetraphosphane **79** is distinguished from its trans isomer (**82**) fundamentally by its reaction with Li(n-Bu). From **71** at 0°C/THF $LiP_4(CMe_3)_2SiMe_3$ (**70**) is formed by the cleavage of a P—Si bond; on the other hand, during the corresponding reaction of **79** between −30 and −25°C in THF, only P—P cleavage with resulting formation of open-chain n-tetraphosphides is observed. The behavior of **79** corresponds therefore to that of the more highly silylated homologs $P_4(SiMe_3)_4$ (**85**) and $P_4(SiMe_3)_3CMe_3$ (**44**). From the reaction of **79** with Li(n-Bu), various n-tetraphosphides can be expected as products of the ring opening. The nucleophilic attack of the butyl moiety occurs only on a silylated phosphorus atom of the four-membered ring [in Scheme 20 on P4], and causes the cleavage of the bond to one of the neighboring atoms P1 or P3. Both possibilities take place, although cleavage of the P4—P1 bond is preferred. According to Scheme 20, the n-tetraphosphides **104** and **105** are formed in a relative proportion of 10:3. The phosphide **103**, which could be thought of as a precursor of **104**, is not demonstrable, even though its formation from **79** and Li(n-Bu) as well as its rapid isomerization into **104**, according to the results given in paragraphs Sections VII,B,1 and VII,B,2, are absolutely plausible. A corresponding isomerization by means of a 1,3-displacement of lithium

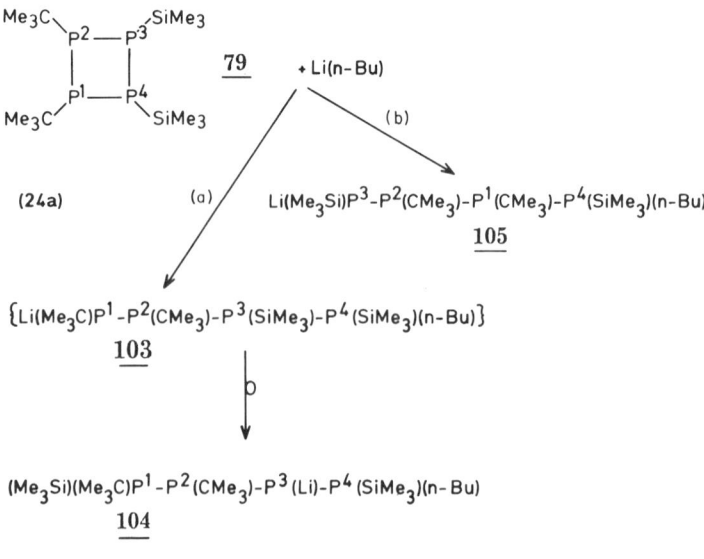

Scheme 20

cannot take place in the primary n-tetraphosphide **105** obtained as indicated in Scheme 20, because the corresponding phosphorus atom is blocked by the nonfunctional CMe_3 group.

Solutions of the n-tetraphosphides **105** and **104** in THF are intensely orange-red. Their decomposition takes place at $-25°C$ within a few hours, and at room temperature within seconds, as shown by the color of the solution turning yellow. The compounds present after a reaction time of 15 days at $20°C$ [molar ratio $79:Li(n\text{-Bu}) = 1:1$] are $LiP_5(CMe_3)_4$ (**76**), $LiP_3(CMe_3)_2$ (**77**), $P_4(CMe_3)_4$ (**110**), $LiP(SiMe_3)_2$, n-$BuP(SiMe_3)_2$ (**111**), and $(Me_3C)P(SiMe_3)_2$. These results do not yet permit the formulation of detailed statements about the course of the reaction which yields the above compounds. Nevertheless they are certainly controlled decomposition reactions of the n-tetraphosphides formed from cis-$P_4(SiMe_3)_2(CMe_3)_2$ (**79**) and $Li(n\text{-Bu})$. And there exist certain analogies with the reactive behavior of $LiP[P(SiMe_3)CMe_3]_2$, which also yields **76** and **77**.

a. Preparation and Properties of the n-Tetraphosphanes from cis-$P_4(SiMe_3)_2(CMe_3)_2$. The thermolabile n-tetraphosphides **105** and **104** described in Section VII,B,3 were reacted with MeCl or Me_3SiCl immediately after their formation from **79** and $Li(n\text{-Bu})$ (44). The phosphides present in the mixture react with MeCl, as shown in Eqs. (19) and (20) between -30 and $-35°C$ within a few minutes to yield the stable n-tetraphosphanes **108** and **109**.

$107 + \text{MeCl} \longrightarrow Me_3C(Me_3Si)P-P(CMe_3)-P(Me)-P(SiMe_3)(n\text{-Bu}) + \text{LiCl}$ (19)

(**108**)

$105 + \text{MeCl} \longrightarrow Me(Me_3Si)P-P(CMe_3)-P(CMe_3)-P(SiMe_3)(n\text{-Bu}) + \text{LiCl}$ (20)

(**109**)

These compounds could not be isolated in pure form. They appear during the final work-up of the reaction solution as viscous liquids which are colorless to light yellow and which mix well under all relative concentrations with the usual solvents and show no tendency whatsoever to crystallize. From the integration of the $^{31}P\{^1H\}$-NMR spectra of these mixtures, a ratio of 10:3 was established for the isomeric compounds **108** and **109**. This agrees with the ratio in the phosphides initially present (44). Although all the n-tetraphosphides so far described can be transformed into the corresponding silyl derivatives by reaction with Me_3SiCl, in the silylation of the phosphide **104** the following peculiarity is encountered.

$$104 \xrightarrow[-\text{LiCl}]{+\text{Me}_3\text{SiCl}} \{\text{Me}_3\text{C}(\text{Me}_3\text{Si})\text{P}-\text{P}(\text{CMe}_3)-\text{P}(\text{SiMe}_3)-\text{P}(\text{SiMe}_3)(n\text{-Bu})\} \quad (21\text{a})$$
$$(110) \downarrow$$
$$\text{P}_3(\text{CMe}_3)_2\text{SiMe}_3 + n\text{-BuP}(\text{SiMe}_3)_2 \quad (21\text{b})$$
$$(111) \qquad (113)$$
$$105 \xrightarrow[-\text{LiCl}]{+\text{Me}_3\text{SiCl}} (\text{Me}_3\text{Si})_2\text{P}-\text{P}(\text{CMe}_3)-\text{P}(\text{CMe}_3)-\text{P}(\text{SiMe}_3)(n\text{-Bu}) \quad (21\text{c})$$
$$(112)$$

After warming the reaction mixture to 20°C the cyclotriphosphane **111** and **113** are formed, instead of the expected n-tetraphosphane **110**. Formation of **111** probably results from 1,3-elimination of n-BuP(SiMe$_3$)$_2$ (**113**) from compound **110**. This preliminary step to the cyclization reaction, shown in Eq. (21b), could not be demonstrated, but nevertheless its formation according to Eq. (21a) and Eq. (19) is plausible. As is also shown by mass spectroscopic investigations, the n-tetraphosphanes exhibit a considerable tendency toward the formation of cyclotriphosphanes by means of 1,3-elimination of simpler phosphanes RP(SiMe)$_2$ (44). The isomeric n-tetraphosphane **112** on the other hand, is stable in the same reaction solution.

ACKNOWLEDGMENTS

I wish to thank my co-workers Dr. J. Härer, Dr. R. Biastoch, Dr. K. Stoll, Dr. T. Vaahs, and P. Amann for their excellent cooperation.

REFERENCES

1. Fritz, G., *Comments Inorg. Chem.* **329** (1982); *Z. Naturforsch. B* **8**, 776 (1953).
2. Fritz, G., *Z. Anorg. Allg. Chem.* **280**, 332 (1955); Fritz, G., and Berkenhoff, H. O., *Z. Anorg. Allg. Chem.* **289**, 250 (1957).
3. Fritz, G., and Emül, R., *Z. Anorg. Allg. Chem.* **416**, 19 (1975).
4. Nöth, H., and Schrägle, W., *Chem. Ber.* **98**, 352 (1965); Drake, J. E., and Simpson, J., *Inorg. Chem.* **6**, 1984 (1967); Drake, J. E., and Goddard, N., *J. Chem. Soc. A*, p. 662 (1969).
5. Becher, H. J., and Langer, E., *Angew. Chem.* **85**, 910 (1973); Issleib, K., and Schmidt, H., *Z. Anorg. Allg. Chem.* **459**, 131 (1979); Kunzek, H., Braun, M., Nesener, E., and Rühlmann, K., *J. Organomet. Chem.* **49**, 149 (1973); Kunzek, H, and Rühlmann, K., *J. Organomet. Chem.* **42**, 391 (1972); Becher, H. J., Fenske, D., and Langer, E., *Chem. Ber.* **106**, 177 (1973); Becker, G., *Z. Anorg. Allg. Chem.* **423**, 242 (1976); **430**, 66 (1977).
6. Appel, R., and Barth, V., *Angew. Chem.* **91**, 497 (1979); *Angew. Chem. Int. Ed. Engl.* **18**, 469 (1979); Appel, R., Barth, V., Halstenberg, M., Huttner, G., and von Seyer, J., *Angew. Chem.* **91**, 935 (1979).
7. Fritz, G., and Poppenburg, G., *Naturwissenschaften* **49**, 449 (1962).

8. Schäfer, H., and MacDiarmid, A. G., *Inorg. Chem.* **15,** 848 (1976).
9. Fritz, G., Schäfer, H., Demuth, R., and Grobe, J., *Z. Anorg. Allg. Chem.* **407,** 287 (1974).
10. Fritz, G., Schäfer, H., and Hölderich, W., *Z. Anorg. Allg. Chem.* **407,** 266 (1974).
11. Fritz, G., Becker, G., and Kummer, D., *Z. Anorg. Allg. Chem.* **372,** 171 (1970); Fritz, G., and Becker, G., *Z. Anorg. Allg. Chem.* **372,** 180 (1970).
12. Fritz, G., and Hölderich, W., *Z. Anorg. Allg. Chem.* **422,** 104 (1976).
13. Fritz, G., and Hölderich, W., *Z. Anorg. Allg. Chem.* **431,** 61 (1977), **431,** 76 (1977).
14. Becker, G., and Hölderich, W., *Chem. Ber.* **108,** 2484 (1975).
15. Fritz, G., and Hölderich, W., *Naturwissenschaften* **62,** 573 (1975); Hönle, W., and von Schnering, H. G., *Z. Anorg. Allg. Chem.* **440,** 171 (1978).
16. Fritz, G., and Uhlmann, R., *Z. Anorg. Allg. Chem.* **440,** 168 (1978).
17. Baudler, M., Ternberger, H., Faber, W., and Hahn, J., *Z. Naturforsch. B. Anorg. Chem. Org. Chem.* **34,** 1690 (1979).
18. Baudler, M., and Faber, W., *Chem. Ber.* **113,** 3394 (1980); Baudler, M., *Angew. Chem.* **94,** 520 (1982).
19. Fritz, G., and Härer, J., *Z. Anorg. Allg. Chem.* **504,** 23 (1983).
20. Fritz, G., Härer, J., and Scheider, K. H., *Z. Anorg. Allg. Chem.* **487,** 44 (1982).
21. Fritz, G., Härer, J., Stoll, K., and Vaahs, T., *Phosphorus Sulfur,* **18,** 65 (1983).
22. Baudler, M., and Exner, O., *Chem. Ber.* **116,** 1268 (1983).
23. Parshall, G. W., and Lindsey, R. V., *J. Am. Chem. Soc.* **81,** 6273 (1959).
24. Schumann, H., and Benda, H., *Chem. Ber.* **104,** 333 (1971).
25. Oakey, R. T., Stanislawski, D. A., and West, R., *J. Organomet. Chem.* **157,** 389 (1978).
26. Schumann, H., and Benda, H., *Angew. Chem.* **81,** 1049 (1969).
27. Fritz, G., and Hölderich, W., *Z. Anorg. Allg. Chem.* **431,** 76 (1977); **475,** 127 (1979); Fritz, G., Uhlmann, R., and Hölderich, W., *Z. Anorg. Allg. Chem.* **442,** 86 (1978).
28. Fritz, G., and Uhlmann, R., *Z. Anorg. Allg. Chem.* **442,** 95 (1978).
29. Hönle, W., and von Schnering, H. G., *Z. Anorg. Allg. Chem.* **442,** 107 (1978).
30. Schäfer, H., Fritz, G., and Hölderich, W., *Z. Anorg. Allg. Chem.* **428,** 222 (1977).
31. Klingebiel, U., and Vater, N., *Angew. Chem.* **94,** 870 (1982).
32. Brauer, G., and Zintel, E., *Z. phys. Chem. Abt. B* **37,** 323 (1937).
33. Issleib, K., and Kümmel, R., *J. Organomet. Chem.* **3,** 84 (1965); Issleib, K., and Tzschach, A., *Chem. Ber.* **92,** 1118 (1959).
34. Fritz, G., and Biastoch, R., *Z. Anorg. Allg. Chem.* **535,** 63 (1986).
35. Fritz, G., Biastoch, R., Hönle, W., and von Schnering, H. G., *Z. Anorg. Allg. Chem.* **535,** 86 (1986).
36. Fritz, G., and Biastoch, R., *Z. Anorg. Allg. Chem.* **535,** 95 (1986).
37. Fritz, G., and Amann, P., *Z. Anorg. Allg. Chem.* **535,** 106 (1986).
38. Fritz, G., and Härer, J., *Z. Anorg. Allg. Chem.* **481,** 185 (1981); Fritz, G., and Stoll, K. *Z. Anorg. Allg. Chem.* **538,** 113 (1986).
39. Fritz, G., and Stoll, K., *Z. Anorg. Allg. Chem.* **538,** 78 (1986).
40. Fritz, G., and Vaahs, T., in preparation.
41. Schumann, H., Rösch, L., and Schmidt-Fritsche, W., *Chem. Ztg.* **101,** 56 (1977).
42. Baudler, M., Hofmann, G., and Hallab, M., *Z. Anorg. Allg. Chem.* **466,** 71 (1980).
43. Fritz, G., and Stoll, K., *Z. Anorg. Allg. Chem.* **514,** 69 (1984).
44. Fritz, G., and Stoll, K., *Z. Anorg. Allg. Chem.* **539,** 65 (1986).
45. Baudler, M., Ternberger, H., Faber, W., and Hahn, J., *Z. Naturforsch. B. Anorg. Chem. Org. Chem.* **346,** 1690 (1979); Baudler, M., Pontzen, T., Hahn, J., Ternberger, H., and Faber, W., *Z. Naturforsch. B. Anorg. Chem. Org. Chem.* **356,** 517 (1980); Baudler, M., and Faber, W., *Chem. Ber.* **113,** 3394 (1980).
46. Fritz, G., Härer, J., and Matern, E., *Z. Anorg. Allg. Chem.* **504,** 38 (1983).

47. Fritz, G., Härer, J., and Stoll, K., *Z. Anorg. Allg. Chem.* **504,** 47 (1983).
48. Baudler, M., Hellmann, J., and Reuschenbach, G., *Z. Anorg. Allg. Chem.* **509,** 38 (1984).
49. Baudler, M., Reuschenbach, G., and Hahn, J., *Z. Anorg. Allg. Chem.* **482,** 27 (1981).
50. Baudler, M., Reuschenbach, G., Hellmann, J., and Hahn, J., *Z. Anorg. Allg. Chem.* **499,** 89 (1983).
51. Baudler, M., Carlsohn, B., Koch, D., and Medda, P. K., *Chem. Ber.* **111,** 121 (1978).
52. Baudler, M., Koch, D., and Carlsohn, B., *Chem. Ber.* **111,** 1217 (1978).
53. Baudler, M., Reuschenbach, G., Koch, D., and Carlsohn, B., *Chem. Ber.* **113,** 1264 (1980).
54. Baudler, M., Reuschenbach, G., and Hahn, J., *Chem. Ber.* **116,** 847 (1983).

INDEX

A

Acetylenes, reactions with polysulfidemetal complexes, 108
Actinide carbides
  metallothermic reduction, 8-10
  reaction with iodine, 10-11
  synthesis, 9
Actinide halides, metallothermic reduction to actinide metal, 4-7
Actinide metals, *see also* specific metals
  availability and price, 2
  chemical properties, 3-4
  crystal growth, 14-15
  early studies, 1-2
  melting point, 7
  physical properties, 3, 36
  preparation, 1-41
    by metallothermic reduction
      of actinide carbides, 8-10
      of actinide halides, 4-6
      of actinide oxides, followed by distillation, 7-8
    by molten salt electrolysis, 11
    by van Arkel-De Boer process, 10-11
  purification methods
    difficulties, 3-4
    electrorefining, 13
    selective vaporization, 12-13
    vacuum melting without distillation, 11-12
    van Arkel-De Boer process, 13
  purity, 3
  radioactivity and toxicity, 3-4
  vapor pressures, 5-6
Actinide oxides
  carbothermic reduction, 9
  metallothermic reduction to actinide metals, 6-8
Actinium
  availability and price, 2
  physical properties, 36
  preparation and purification, 7, 16-17
  radioactive decay, 16
Actinium carbides, 17
Actinium halides, metallothermic reduction, 16
Actinium oxides, metallothermic reduction, 7, 16
Activation energies, for nucleophilic astatination of halobenzenes, 59
Aldehydes and ketones, reactions with iminoboranes, 159, 160, 162
Alkyl astatides, 53-55
  physicochemical properties, 54
Alkylating reagents, addition to iminoboranes, 157
Alkynes
  addition to iminoboranes, 163
  in transition metal complexes, 165
Allyloboration, of iminoboranes, 156-157
Americium
  availability and price, 2
  crystal growth, 14-15
  distilled metal, 30
  melting point, 6
  oxide reduction-metal distillation still, 29
  physical properties, 36
  preparation and purification, 5, 7-8, 12-13, 26-28
    apparatus, 29
  purity, 3
  radioactivity, 26-27
  vapor pressure, 6-7
Americium-241, 26-27
  neutron irradiation of, 24
  radioactive decay, 21
Americium dioxide, metallothermic reduction, 7, 28
  rate and yield, 8
Americium trifluoride, metallothermic reduction, 27
Amino acids, aromatic, astatination, 67-69
Aminoboranes
  reaction to form borazines, 127-128
  synthesis, 151
Aminoboration, of iminoboranes, 155
5-Aminouracil, astatination, 75

INDEX

Astatinated amino acids and proteins, 67–72
Astatinated nucleosides and nucleotides, 75–77
Astatine, 43–88
  α-particle energy, 45
  biochemical fate, 78
  biological behavior, 77–78
  biomedical applications, 79–83
    therapeutic studies, 80–81
  diatomic, 50
  distallation, 47–48
  embryotoxicity, 78
  extraction techniques, 47
  identification, 49
  isotopes, 43–49
    decay, 44
    half-lives, 44
  monovalent, compounds, 53–77
  multivalent, compounds, 52–53
  preparation, 44–49
    excitation functions, 46
    by nuclear reactions, 45–48
    by secondary nuclear reactions and spallation, 48
  radioactivity, 43–44, 50, 79
  synthetic organic radiochemistry, 49–77
    criteria and guidelines, 51–52
    limitations, 49–52
    reaction reversibility, 50
  uptake in thyroid, 77–78
Astatine-211, decay scheme, 79
Astatoanilines, 65–66
Astatobenzene, 55–60
  physicochemical properties, 62
  synthesis, 56–59
Astatobenzoic acids, 67
  dissociation constant, Hammett σ-constant, field, and resonance effects, 66
Astatocarboxylic acids, 55
6-Astatocholesterol, 74
2-Astatoestradiol, 74
4-Astatoestradiol, 74
Astatohalobenzenes, 61–64
  physicochemical properties, 62–63
Astatoimidazoles, 74–75
2-Astato-4-iodoestradiol, 74
3-Astato-5-iodotyrosine, 68
3-Astato-4-methoxyphenylalanine, 68
6-Astato-2-methyl-1,4-naphthoquinol diphosphate, synthesis, 72–73
  biomedical applications in cancer therapy, 81
6-Astatomethyl-19-norcholest-5(10)-en-3β-ol, 74
Astatonaphthoquinones, 72–73
  analysis by thin-layer chromatography, 72
Astatonitrobenzenes, 66–67
  physicochemical properties, 62–63
Astatophenazathioniums, 77
  biomedical applications in cancer therapy, 81
Astatophenols, 64–65
  dissociation constants, Hammett σ-constants, field, and resonance effects, 66
4-Astatophenylalanine, 68
Astatopyrimidines, 75–77
Astatosteroids, 73–74
Astatotoluenes, 60–61
  physicochemical properties, 62
3-Astatotyrosine, 68
Astatouracil, 75
  biomedical applications, 81
5-Astatouridine, 75–76
Atomic volume, of actinide metals, 36
Azides, addition to iminoboranes, 163
Azidoboranes, decomposition, 125, 128–129, 132–133
Azidoboration, of iminoboranes, 154–155
Azidosilation, of iminoboranes, 158, 163

B

Barium, in metallothermic reduction of actinide halides, 5
Berkelium
  availability and price, 2
  melting point, 6
  physical properties, 36
  preparation and purification, 7, 12, 31–33
    apparatus, 32
  purity, 2
  radioactivity, 32
  vapor pressure, 6
Berkelium oxides, reduction by thorium, 7
Bidentate polysulfide complexes, bond lengths, 114

Binuclear polysulfidemetal complexes, 99–101
Biomedical applications, of astatine, 79–83
Bismuth-209, α-particle bombardment, 46, 49
Bismuth, polysulfide complex, 100
  boat chair conformation, 115
Boat chair conformation, in polysulfidemetal complexes, 115
Boiling points, of organoastatine compounds, 54, 62–63
Bonding orbitals, in disulfidemetal complexes, 113
Bond lengths
  in bidentate polysulfide complexes of metals, 114
  in coordinated polysulfidemetal complexes, 94
  in dioxygen and disulfur, 112
  in iminoborane cyclodimers, 144
Borazine
  analogy to benzene, 146
  fluxional behavior, 146
Borazines, 146–147
  synthesis, 127–128, 149
t-Butyldichlorophosphine, reactions with lithiated silylphosphanes, 184–185
n-Butyllithium, reactions with cis-$P_4(SiMe_3)_2(CMe_3)_2$, 210–212

C

Calcium, in metallothermic reduction of actinide halides and oxides, 5–7
Californium
  availability and price, 2
  melting point, 6
  physical properties, 36
  preparation and purification, 5, 7, 12, 33
    apparatus, 34, 35
  purity, 3
  radioactivity, 33
  vapor pressure, 6
Californium-252, α-decay, 28
Californium oxide, metallothermic reduction, 7, 33
  apparatus, 34, 35
Californium trifluoride, metallothermic reduction, 33
Cancer therapy, astatine-211 in, 79–83
Carbothermic reduction, of actinide oxides, 9
Chair conformation, in polysulfidemetal complexes, 115
Chloroboration, of iminoboranes, 153–154
Cholesterol, astatination, 7
Chromium
  iminoborane complex, 166
  polysulfide complexes, 96, 98
Clusters, 96
Cobalt
  iminoborane complex, 165
  polysulfide complexes, 96
Condensed ring systems, in polysulfidemetal complexes, 101–102
Conformations, of polysulfidemetal ring complexes, 115
Coordination, in polysulfide ligands, 91–97
  end-on coordination, 91–93, 95–97
  side-on coordination, 91–95
Copper, polysulfide complexes, 100, 101
  envelope conformation, 115
Cradle conformation, in polysulfidemetal complexes, 100
Crown conformation, in polysulfidemetal complexes, 115
Crystal growth, of actinide metals, 14–15
Crystal structure, of actinide metals, 36
Curium
  availability and price, 2
  crystal growth, 14–15
  distilled metal, 31
  isotopes, 28
  physical properties, 36
  preparation and purification, 7, 10, 12–13, 29–31
  purity, 3
  radioactivity, 28–29
Curium fluorides, metallothermic reduction, 29–31
Curium oxides, metallothermic reduction, 7, 30–31
Cyclic iminoboranes, as reaction intermediates, 130
Cyclic silylphosphanes, see Silylphosphanes, phosphorus-rich, cyclic
Cycloaddition reactions, of iminoboranes, 159–165
Cyclobutadienes, 144, 166

Cyclochloroboranes
  amination, 131
  reaction with trimethylsilyl azide, 130
Cyclodimers, of iminoboranes, see Iminoboranes, cyclodimers
Cyclooctatetraene, 147
Cyclopentadiene, addition to iminoborane, 164

D

Density, of actinide metals, 36
Dewar benzenes, 146-147
Dewar borazines, 146-147
Diarylazidoboranes, decomposition, 129
Diazadiboretidines, 143-146
Diborylamines, synthesis, 128-129
Di-t-butyldichlorosilane, reactions with lithium phosphides, 181-186
Dichlorodiethylsilane, reactions with lithium phosphides, 179-180
Dichlorodimethylsilane, reaction with lithium phosphides, 177-179, 185, 187
Dioxygen, bond lengths and vibrational frequencies, 112
Direct oxide reduction, in actinide metal preparation, 6-7, 21-22, 25
  apparatus, 29
Disproportionation, of astatine halides, 50-51
Dissociation constants, of astatophenols, 66
Distillation
  of actinides, 7-8
  of astatine, 47-48
Disulfido metal complexes
  bonding orbitals, 113
  spectroscopy, 109-110
  structure and bonding, 111-113
Disulfur, bond lengths and vibrational frequencies, 112

E

Einsteinium
  melting point, 6
  physical properties, 36
  preparation and purification, 5, 7, 34-36
  purity, 3
  radioactivity, 34
  vapor pressure, 6

Einsteinium oxide, reduction by lanthanum, 7, 35
Einsteinium trifluoride, metallothermic reduction, 34-35
Electrolysis, of neptunium(III), 21
Electron transfer reactions, of polysulfidemetal complexes, 106
Electrorefining, of actinide metals, 13, 15, 17, 20, 21
  deposition of neptunium crystals, 21
  plutonium crystals, 26
Enthalpy of vaporization
  of actinide metals, 36
  of aliphatic astatides, 54
  of aromatic astatides, 62-63
Envelope conformation, in polysulfidemetal complexes, 98, 102, 115
Estradiol, astatination, 73-74
Excitation functions, for production of astatine, 46

F

Fermium, 4
Field effects, of astatophenols, 66

G

Geometries, of polysulfidemetal complexes, 92
Gold, polysulfide complex, 100
  ring conformation, 115
Grain coarsening, 14

H

Half chair conformation, in polysulfidemetal complexes, 98, 115
Halobenzenes, astatination, 56-59
  activation energies, 59
Halosilanes, addition to iminoboranes, 157-158
Hammett σ-constants, of astatophenols, 66
Heteroallenes, cycloaddition to iminoboranes, 161
Heterocyclic astatides, 74-77

I

Imidazoles, astatination, 74-75
Iminoalkanes, cycloaddition to iminoboranes, 160-161

Iminoborane, atomic charges, 134
Iminoboranes, 123-170
  bonding, 133-134
  bond lengths, 135-137
    compared to organic compounds, 137
  in cycloaddition reactions, 159-165
    with aldehydes and ketones, 160, 162
    with alkynes, 163
    with azides, 163-164
    [2 + 2]-cycloadditions, 159-163
    [3 + 2]-cycloadditions, 163-164
    [4 + 2]-cycloadditions, 164-165
    with cyclopentadiene, 164
    with heteroallenes, 161
    with iminoalkanes, 160-161
    with iminophosphanes, 162
    with nitrones, 164
    with phosphaalkenes and -alkynes, 163
  cyclodimers, 143-146
    ring bond lengths and angles, 144
  cyclotetramers, 147-148
  cyclotrimers, 146-147
  dipole moments, 139
  formation, 124-133
    of cyclooligomers, 129
    of metastable species, 124-127
    of symmetric iminoboranes, 125
  isolated species, 126
  kinetic stability, 127, 134
  oligomerization, 141-150
    cyclooligomer interconversions, 148-150
  polar additions to
    of aldehydes and ketones, 159
    of alkylation reagents, 157
    allyloboration, 156-157
    aminoboration, 155
    azidoboration, 154-155
    to both π-bonds, 158-159
    chloroboration, 153-154
    of halosilanes, 157-158
    of Lewis acids or bases, 150-151
    organoboration, 155-156
    of protic agents, 152
    thioboration, 155
  as reaction intermediates, 127-132
    intramolecular stabilization, 131-132
    stabilization products, 141-143
  spectroscopy, 125-126, 137-139
  structural formulas, 140
  structure, 133-140
  thermodynamic stability, 133-134
  in transition metal complexes, 165-167
  X-ray structural studies, 135
Iminophosphanes, cycloaddition to iminoboranes, 162
Iodide transport, *see* van Arkel-De Boer process
3-Iodotyrosine, astatination, 67-68
Ionization potentials, of organoastatine compounds, 54, 62-63
Iridium, polysulfide complexes, 93
Iron, mixed-metal polysulfide complex, 95
Iron, polysulfide complexes, synthesis, 104-105

L

Lanthanum, in metallothermic reduction of actinide oxides, 7-8
Lawrencium, 4
Lewis acids and bases, addition to iminoboranes, 150-151
Ligand migration, in polysulfidemetal complexes, 108
Lithium, in metallothermic reduction of actinide halides, 5
Lithium alkyls, in silylphosphane chemistry, 172-174, 192
  reactions with $(Me_3C)(Me_3Si)P-PCl_2$, 190-191
  reactions with $[(Me_3C)(Me_3Si)P]_2PLi$, 201-204
  reactions with $(Me_3Si)_2P-P(Li)-P-SiMe_3)(CMe_3)$, 199-201
  reactions with $P_4(SiMe_3)_4$, 205-207
  reactions with $P_4(SiMe_3)_3CMe_3$, 207-209
Lithium phosphides
  Li_2P_{14}, synthesis and reactions, 174-175
  Li_3P_7, synthesis and reactions, 174-175
  reactions
    with t-butyldichlorophosphine, 184-185
    with di-t-butyldichlorosilane, 181-186
    with dichlorodiethylsilane, 179-180
    with dichlorodimethylsilane, 177-178
    with halosilanes, 177
    with $HP[Si(CMe_3)_2]PH$, 181-184

Lithium phosphides, reactions, *continued*
  with $(Me_3C)(Me_3Si)P—PCl_2$, 190-191
  with silylated triphosphanes and cyclotetraphosphanes, 199-212
  in silylphosphane chemistry, 172, 174-212

M

Manganese, polysulfide complex, 95
Melting point, of actinide metals, 36
Mendelevium, 4
Mercury, polysulfide complex, 98
  twist chair conformation, 115
Metallothermic reduction
  of actinide carbides, 8-10
  of actinide halides, 4-6
  of actinide oxides, 6-8
  of actinium halides, 16
  of actinium oxides, 16
  of americium oxide, 27-28
  of americium trifluoride, 27
  of berkelium fluorides, 32
    reaction apparatus, 32
  of californium oxide, 33
    apparatus, 34, 35
  of californium trifluoride, 33
  of curium fluorides, 29-31
  of curium oxides, 7, 30-31
  of neptunium fluorides, 21
  of plutonium carbide, 26
  of plutonium oxides, 7, 25
  of plutonium tetrafluoride, 24
  of protactinium tetrafluoride, 18
  of thorium tetrachloride, 17
  of uranium halides, 20
Metastable iminoboranes, synthesis, 124-127
4-Methoxyphenylalanine, astatination, 67-68
Methylene blue, astatination, 77
Molten salt electrolysis, 11
Molybdenum, polysulfide complexes, 93, 95-97, 100
  crown conformation, 115
  reactions, 106-108
  synthesis, 103-105
Mononuclear polysulfidemetal ring complexes, 97-99

N

Neptunium
  analysis of ultrapure metal, 22
  availability and price, 2
  crystal growth, 14
  electrodeposited metal, 21
  impurities, 22
  melting point, 6
  physical properties, 36
  preparation and purification, 5, 7, 11, 13, 21-23
  purity, 3
  radioactivity, 21
  vapor pressure, 6
Neptunium fluorides, metallothermic reduction, 21
Nickel, polysulfide complex, 98
  half chair conformation, 115
Niobium, polysulfide complex, 100
  boat chair conformation, 115
Nitrones, cycloaddition to iminoboranes, 164
Nobelium, 4
Nuclear reactor fuel, isotopic composition, 23
Nucleic acids, astatination, 76-77
Nucleophiles, reactions with polysulfide-metal complexes, 106-107
Nucleosides and nucleotides, astatination, 75-77

O

Oligomerization, of iminoboranes, *see* Iminoboranes, oligomerization
Organoastatine compounds, identification, 51, *see also* specific compounds and classes of compounds
Organoboration, of iminoboranes, 155-156
Osmium, polysulfide complexes, 100
Oxidation, of polysulfidemetal complexes, 107
Oxide reduction-metal distillation still apparatus, 29

P

Palladium, polysulfide complexes, 100
  synthesis, 103

Phenylalanine, astatination, 67–68
Phosphaalkenes and -alkynes, cycloaddition to iminoboranes, 163
Phosphorus-rich silylphosphanes, *see* Silylphosphanes, phosphorus-rich
Phosphorus trichloride, reactions with silylphosphanes, 189–190
Physical vapor transport, 14
Platinum, polysulfide complexes, 98
  synthesis, 103, 105
Plutonium
  availability and price, 2
  crystal growth, 14
  impurities, 27
  isotopic composition, 23
  melting point, 6
  neutron emission rates, 24
  physical properties, 36
  preparation and purification, 5, 7, 9–13, 24–26
  purity, 3
  ultrapure metal, 26
  vapor pressure, 5–6
Plutonium-239, neutron capture, 28
Plutonium-241, β-decay, 26
Plutonium carbide, tantalothermic reduction, 9, 26
Plutonium oxides, metallothermic reduction, 7, 25
Plutonium tetrafluoride, metallothermic reduction, 24–25
Polynuclear polysulfidemetal complexes, 101–102
Polysulfide ions, 89–91
  as bidentate ligand, 97–99
  bond lengths, 114
  as doubly bridging ligand, 99–101
  as polydentate ligand, 101–102
  preparation, 91
  stereochemistry, 90
Polysulfidemetal complexes, 89–122
  cluster structures, 96
  compound types, 91–103
  with coordinated polysulfide ligands, 91–97
    cage structures, 96
    end-on coordinated (type II) complexes, 95–96
    side-on coordinated (type I) complexes, 93–95
    sulfur–sulfur bond lengths, 94
    type III complexes, 96–97
    vibration frequencies, 94
  geometries, 92
  reactions, 106–108
    with acetylenes, 108
    electron transfer, 106
    ligand migration, 108
    with nucleophiles, 106–107
    oxidation of sulfur ligand, 107
    ring cleavage, 108
    thermal decomposition, 107–108
  ring conformations, 115
  solid-state structures, 103
  spectroscopic properties, 109–111
  structure and bonding, 111–116
  with sulfur rings, 97–103
    binuclear, 99–101
    with mixed polysulfido ligand, 103
    mononuclear, 97–99
    polynuclear, 101–102
    with strong metal–metal bonds, 102–103
  synthesis, 103–105
Preparation, of actinide metals, *see* Actinide metals, preparation
Protactinium
  availability and price, 1
  crystal growth, 14
  crystals, 15, 18
  melting point, 6
  physical properties, 36
  preparation and purification, 10, 11, 13, 17–19
  purity, 3
  van Arkel–De Boer purification, 19
  vapor pressure, 6
Protactinium carbide, in van Arkel–De Boer process, 18
Protactinium pentahalide, thermal decomposition, 18
Protactinium tetrafluoride, metallothermic reduction, 18
Proteins, astatination, 68–72
Protic agents, addition to iminoboranes, 152
Purification, of actinide metals, *see* Actinide metals, purification

R

Radioactivity
  of actinium, 16
  of americium, 26–27
  of astatine, 43–45, 79
  of berkelium, 32
  of californium, 33
  of einsteinium, 34
  of neptunium, 21
  of plutonium, 23–24
  of thorium, 17
  of uranium, 19
Radon-211, in preparation of astatine, 49
Radon-226, neutron bombardment, 16
Recrystallization, of actinide metals, 14
Reduction
  of actinide carbides, 8–10
  of actinide halides, 4–6
  of actinide oxides, 6–8
Refining, of actinide metals, see Actinide metals, purification
Resistivity ratio, 13
Resonance effects, of astatophenols, 66
Rhenium, polysulfide complexes, 100, 102
  envelope conformation, 115
  synthesis, 103–104
Rhodium, polysulfide complexes, synthesis, 103
Ring systems, in polysulfidemetal complexes, 97–103
  cleavage, 108
  condensed, 101–102
  conformations, 115
  with strong metal–metal bonds, 102–103
  structural types, 97

S

Selective vaporization, for preparation of actinide metals, 12–13, 26
Silver, polysulfide complexes, 98, 101
  ring conformations, 115
Silylated cyclotetraphosphanes, 198
  reactions with lithium alkyls and lithium phosphides, 204–212
Silylated tri- and tetraphosphanes
  reactions with lithium alkyls and lithium phosphides, 199–212
  synthesis via lithiated diphosphanes, 194–198
Silylated triphosphanes and triphosphides, synthesis, 188–194
  yields, 194
Silylphosphanes, phosphorus-rich, 171–214
  cyclic, synthesis, 175–188
  $HP[Si(CMe_3)_2]PH$, 181–184
  $P_4(SiMe_2)_6$, 176
  $P(SiMe_2)_3P$, 185–186
  by reaction of lithium phosphides with di-t-butyldichlorosilane, 181–186
  by reaction of lithium phosphides with dichlorodiethylsilane, 179–180
  by reaction of lithium phosphides with dichlorodimethylsilane, 177–179
  lithiated, reactions 174, 184–185
  $(Me_3C)(Me_3Si)P—PCl_2$, reactions with lithium alkyls, 190–191
  $[(Me_3C)(Me_3Si)P]_2PLi$, 192
    reactions with lithium alkyls, 201–204
  $(Me_3C)P(SiMe_3)_2$, reaction with phosphorus trichloride, 189–190
  $MeP(SiMe_3)_2$, reaction with phosphorus trichloride, 189–190
  $(Me_3Si)_2P—P(H)—P(SiMe_3)CMe_3$, 191–192
  $(Me_3Si)_2P—P(Li)—P(SiMe_3)CMe_3$, 192
    reactions with lithium alkyls, 199–201
  $P_4(SiMe_2)_3$, 173
  $P_4(SiMe_3)_4$, reactions with lithium alkyls, 205–207
  $P_7(SiMe_3)_3$
    formation, 173–175
    structure, 173
  cis-$P_4(SiMe_3)_2(CMe_3)_2$, reactions with lithium alkyls, 210–212
  $P_4(SiMe_3)_3CMe_3$, reactions with lithium alkyls, 207–209
  synthesis, 173–188
[Silyl(silyloxy)amino]boranes, decomposition, 129
Solid-state electrotransport, for purification of actinide metals, 13, 14
Solid-state structures, of polysulfidemetal complexes, 103
Spectroscopy, of polysulfidemetal complexes, 109–111
SSEP, see Solid-state electrotransport
Stabilization products, of iminoboranes, 141–143

Stereochemistry, of polysulfide ions, 90
Steroids, astatination, 73-74
Strain annealing, 14
Sulfur ligands, *see* Polysulfide ions
Synthesis
  of actinide carbides, 9
  of actinide metal crystals, 14-15
  of alkyl astatides, 53
  of aminoboranes, 151
  of astatinated amino acids and proteins, 67-72
  of astatoanilines, 65-66
  of astatobenzene, 56
  of astatocarboxylic acids, 55
  of astatohalobenzenes, 61, 64
  of astatonaphthoquinones, 72-73
  of astatonitrobenzenes, 67
  of astatophenols, 64-65
  of astatosteroids, 73-74
  of astatotoluenes, 60-61
  of borazines, 127-128, 149
  of cyclic silylphosphanes, 175-188
  of η^4-diazadiboretidinemetal compounds, 166
  of diborylamines, 128-129
  of hydrazones, 132
  of iminoboranes, 124-133
  of Li_2P_{14}, 174-175
  of Li_3P_7, 174
  of metastable iminoboranes, 124-127
  of phosphorus-rich silylphosphanes, 173-188
  of polysulfide ions, 91
  of polysulfidemetal complexes, 103-105
  of silylated tri- and tetraphosphanes, 194-198
  of silylated triphosphanes and triphosphides, 188-194
  of *n*-tetraphosphanes, 206-212

T

Tantalum
  distillation columns, for purification of actinide metals, 12
  reduction of plutonium carbide, 9
n-Tetraphosphanes, 206-212
Thermal decomposition of polysulfidemetal complexes, 107-108

Thin-layer chromatography, of astatonaphthoquinones, 72
Thioboration, of iminoboranes, 155
Thorium
  availability and price, 2
  crytal growth methods, 14-15
  metallothermic reduction of actinide oxides, 7-8
  physical properties, 36
  preparation and purification, 5, 10, 11, 13, 17
  purity, 3
  vapor pressure, 5-6
Thorium carbide, reaction with iodine, 10
Thorium tetrachloride, metallothermic reduction, 17
Titanium
  iminoborane complex, 167
  polysulfide complexes, 98, 100, 102
    chair conformation, 115
    reactions, 107-108
    synthesis, 103, 104
Transeinsteinium actinides, 4
Transition metals
  alkyne complexes, 165
  iminoborane complexes, 165-167
  in metallothermic reduction of actinide carbides, 8-10
Trapping agents, in iminoborane chemistry, 128-131
Trialkylboranes, as trapping agents, 128-130
Tungsten, polysulfide complex, 95
  synthesis, 103
Twist chair conformation, in polysulfidemetal complexes, 115
Tyrosine, astatination, 67-68

U

Uranium
  abundance, 19
  availability and price, 2
  average analysis, 20
  crystal growth, 14-15
  impurities, 20
  melting point, 6
  physical properties, 36
  preparation and purification, 5, 7, 11, 13, 19-21

Uranium, *continued*
  purity, 3
  radioactivity, 19
  vapor pressure, 5-6
Uranium-235, radioactive decay, 16, 17, 21
Uranium-238, radioactive decay, 17, 19, 21
Uranium halides, metallothermic reduction, 20

V

Vacuum melting, in actinide metal purification, 11-12, 18, 20, 22, 28

van Arkel-De Boer process, for actinide metal preparation, 10-11, 13, 14, 18-19
  crystals, 15

Z

Zone melting, of actinide metals, 13